SIMULTANEOUS SULFATE REDUCTION AND METAL PRECIPITATION IN AN INVERSE FLUIDIZED BED REACTOR

Denys Kristalia Villa Gómez

Thesis committee

Promotor

Prof. Dr P.N.L. Lens

Professor of Environmental Biotechnology

UNESCO-IHE, Delft

Co-promotor

Dr K.J. Keesman

Associate professor, Biomass Refinery and Process Dynamics

Wageningen University

Other members

Prof. Dr C.J.N. Buisman, Wageningen University

Prof. J. Puhakka, Tampere University of Technology, Finland

Prof. A. Lewis, University of Cape Town, South Africa

Dr J. Huisman, Paques B.V., Balk, The Netherlands

This research was conducted under the auspices of the SENSE Research School for Socio-Economic and Natural Sciences of the Environment

Simultaneous sulfate reduction and metal precipitation in an inverse fluidized bed reactor

Thesis

submitted in fulfilment of the requirements of

the Academic Board of Wageningen University and

the Academic Board of the UNESCO-IHE Institute for Water Education

for the degree of doctor

to be defended in public

on Friday, 18 October 2013 at 4 p.m.

in Delft, the Netherlands

by

Denys Kristalia Villa Gómez
Born in Zacatecas, Mexico

CRC Press/Balkema is an imprint of the Taylor & Francis Group, an informa business

© 2013, Denys K. Villa Gómez

Published by:
CRC Press/Balkema
PO Box 11320, 2301 EH Leiden, The Netherlands
e-mail: Pub.NL@taylorandfrancis.com
www.crcpress.com – www.taylorandfrancis.com

ISBN 978-1-138-00166-4 (Taylor & Francis Group)
ISBN 978-94-6173-741-0 (Wageningen University)

Table of contents

CHAPTER

1

GENERAL
INTRODUCTION

1.1 Preface

The exploitation of minerals has contributed to the economic growth of many regions all over the world. In 1999, for example, Mexico produced 182202 tons of lead (Pb), 1485 tons of cadmium (Cd), 447948 tons of Zn and 402430 of copper (Cu) (www.inegi.gob.mx). Mining activities result in pollution problems due to the emissions of acid mine drainage (AMD) into the environment. AMD is one of the most widespread causes of pollution in the world (Gazea et al. 1996; Johnson and Hallberg 2005). The principles governing the generation of AMD are relatively well understood. Upon exposure to oxygen and water and in the presence of oxidizing bacteria, pyrite and other sulfide minerals are oxidized to produce a leachate containing dissolved metals, sulfate and acidity (Tabak et al. 2003).

1.2 Agreements and legislation

In developed countries, like, Europe, USA and Australia, the source and amount of AMD generated as well as the damage caused to the environment are well described (GAO 1996; Harries 1997). In most developing countries these issues have not been well documented. Information on mining impact in these regions is limited and scattered widely in private and official files. The basic sources are reports produced by large mining firms, environmental organizations and government institutions which deal with environmental protection. As a consequence of this lack of information, the responsibility for the treatment of the wastes generated by the mining activities is also not clear. Small-scale and even some large-scale mining industries in developing countries seldom have their effluents properly treated before discharge into water bodies.

More stringent legislation promoted by international agreements resulted in an increased interest in cost-effective technologies for the treatment of AMD. One example of these international agreements is the partnership for acid drainage remediation in Europe (PADRE), which promotes international best practice in the stewardship of waters and soils on European sites subject to the generation and migration of acidic drainage. Another example of these international agreements is the Commission for Environmental Cooperation (CEC) which is an international organization created by Canada, Mexico and the United States under the North American Agreement on Environmental Cooperation (NAAEC). This commission leads to a regional cooperation for the management of a range of chemical substances of mutual concern; including pollution prevention, source reduction and pollution control.

1.3 Treatment of AMD

1.3.1 Physico-chemical process

Most industries treat metal containing wastewaters such as AMD by precipitation with hydroxide or limestone because of process simplicity, low cost chemicals and easy process control (Fu and Wang 2011). These chemical precipitation processes have drawbacks in terms of application and effectiveness, e.g. hydroxide precipitation usually results in the production of unstable metal hydroxides, which is hardly suitable for metal recovery, leading to a greater disposal expense (Esposito et al. 2006). Other technologies such as ion exchange, reverse osmosis and electro-dialysis are available to treat AMD but they are expensive and not commonly used (Fu and Wang 2011). All these methods require high capital investments and have sludge disposal problems due to the need of more drying facilities and further treatment (Aziz et al. 2008).

Sulfide precipitation is an alternative over the conventional methods such as hydroxide precipitation for several reasons: 1) the formation of highly insoluble salts even in the acid region below pH 7 (Brooks 1991); 2) selective metal recovery due to the different solubility products of the different metal sulfides over the pH region (Sampaio et al., 2009); 3) better settling, thickening and dewatering characteristics than hydroxide precipitates (Lewis and Swartbooi 2006; Esposito et al. 2006; Djedidi et al. 2009) and 4) many metal refining operations, notably for copper and zinc, are designed for processing sulfide ores (Brooks 1991).

1.3.2 Sulfate reducing process

Biogenic sulfide produced by sulfate reducing bacteria (SRB) is also an option for metal sulfide precipitation from wastewater, especially when besides metals, also sulfate is a major wastewater constituent (Equation 2), e.g. in wastewaters from metal refineries and AMD. The biological sulfate reducing process requires the supply of an electron donor and carbon source as well as essential nutrients to maintain microbial activity:

$$2\ CH_2O + SO_4^{2-} \rightarrow H_2S + 2\ HCO_3^-$$

(1)

where CH_2O = electron donor

Removal of metals by SRB is mainly due to the production of highly insoluble precipitates with biogenic sulfide as shown:

$$H_2S + M^{2+} \rightarrow MS_{(s)} + 2H^+ \tag{2}$$

where M^{2+} = metal, such as Zn^{2+}, Cu^{2+}, Pb^{2+} and Ni^{2+}

SRB are either heterotrophic (using organic compounds) or autotrophic (using hydrogen as an electron donor and CO_2 as carbon source) anaerobes, capable of reducing sulfate to sulfide by a dissimilatory, bioenergetic metabolism (Nagpal et al. 2000). SRB are prokaryotes found nearly everywhere in waters and sediments because of their ability to use a wide range of substrates and the ability of many SRB to tolerate extreme conditions (Schwartz 1985).

SRB have optimal environmental requirements which must be met to enable sulfate reducing activity, such as an anaerobic environment (redox potential below -200 mV is generally needed), pH values greater than 5, the presence of an organic substrate or H_2 to be oxidized as electron donor and the presence of an appropriate sulfur species (as sulfate) to be reduced (Gibert et al. 2002).

SRB can, in general, be divided into two major groups (Widdel 1988): those that oxidize the carbon source completely to CO_2 (i.e. *Desulfobacter, Desulfococcus, Desulfosarcina, Desulfonema and Desulfobacterium*) and those that oxidize the carbon source incompletely to acetate, CO_2 and H_2 (i.e. *Desulfotomaculum, Desulfovibrio, Desulfomonas, Desulfobulbus and Termodesulfobacterium*)

Although some SRB are capable of fermentative and sulfidogenic growth on sugars and amino-acids, in anaerobic reactors and in the presence of sulfate, SRB are more likely to be involved in the last stages of mineralization rather than in the initial fermentative stage (Figure 1.1).

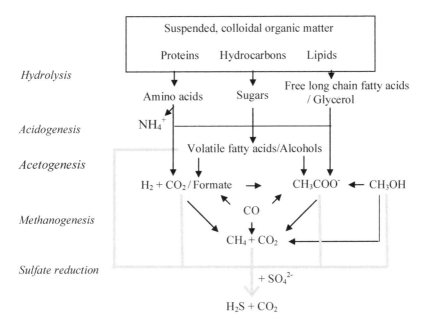

Figure 1.1 Simplified schematic representation of the anaerobic degradation process in the absence (in black) and in the presence (in grey) of sulfate (Bijmans et al. 2011).

1.3.2.1 Electron donor and carbon source for sulfate reduction

Generally AMD contains low concentrations of dissolved organic carbon. For this reason, the addition of organic carbon source (electron donor) determines the overall costs of the sulfate reduction process (Gibert et al. 2004; Zagury et al. 2006). The choice of the substrate is based on different criteria: *i)* the ability of SRB to utilize the organic substrate, *ii)* the sulfate load to be reduced and the cost of the substrate per unit of sulfide produced, *iii)* the availability in sufficient quantities and *iv)* the remaining pollution load in case of incompletely substrate degradation (van Houten et al. 1994; Dries et al. 1998; Dijkman et al. 1999). Electron donor loss because of methane production or acetate production will add to the total costs of the process, and vary per electron donor. It is also important to consider the possible toxicity of intermediate compounds from the incompletely degraded substrate (Kaksonen and Puhakka 2007).

Simple organic carbon sources

SRB use the easily degradable fraction of organic matter such as low molecular weight compounds with simple structures, e.g., methanol, ethanol, lactate (Dvorak et al. 1992;

Nagpal et al. 2000; Tsukamoto et al. 2004), polylactic acid (Edenborn 2004) and simple carbohydrate monomers, e.g., glucose or sucrose (Mizuno et al. 1998).

A higher negative Gibb's free energy change of the lactate oxidation reaction further favors the preference of SRB for lactate as the electron donor and carbon source (Table 1.1). Lactate is also preferable in terms of biomass production; the reason is that it is oxidized either completely or incompletely in the presence of sulfate by a diverse range of SRB strains (Okabe et al. 1995; Kaksonen et al. 2003b) which promotes microbial diversity. The main drawbacks are the high cost on large-scale and the predominance of undissociated lactate in acidic waters such as AMD which could be inhibitory or lethal to SRB (Brüser et al. 2000). With respect to the free energy change, ethanol is the second favorable electron donor; it is also cheaper than lactate (Nagpal et al. 2000). The ability of both incomplete and complete oxidizing SRB to grow with ethanol as electron donor has been demonstrated in several studies (Widdel 1988; Nagpal et al. 2000; Kaksonen et al. 2004).

Table 1.1 Free energy of sulfate reducing reaction with different electron donors (J.W.H et al. 1994).

Equation	$\Delta G°$ (kJ/mol)
$4H_2 + SO_4^{2-} + H^+ \rightarrow HS^- + 4H_2O$	-38.1
Acetate $+ SO_4^{2-} \rightarrow 2HCO_3^- + HS-$	-47.6
Propyonate $+ \frac{3}{4} SO_4^{2-} \rightarrow$ Acetate $+ HCO_3^- + \frac{3}{4} HS^- + \frac{1}{4} H^+$	-37.7
Butyrate $+ \frac{1}{2} SO_4^{2-} \rightarrow 2$ Acetate $+ \frac{1}{2} HS^- + \frac{1}{2} H^+$	-27.8
Lactate $+ \frac{1}{2} SO_4^{2-} \rightarrow$ Acetate $+ HCO_3^- + \frac{1}{2} HS^- + H^+$	-80.0
Ethanol $+ \frac{1}{2} SO_4^{2-} \rightarrow$ Acetate $+ HCO_3^- + \frac{1}{2} HS^- + \frac{1}{2} H^+ + H_2O$	-66.4

Complex organic carbon sources

Alternatively, less expensive organic carbon sources such as waste materials from the agricultural and food processing industry have been assessed for their potential to sustain sulfate reduction. Various organic wastes have been used as electron donor for the SRB applied in the treatment of AMD including straw and hay (Bechard et al. 1994), oak chips (Chang et al. 2000), spent mushroom compost (Dvorak et al. 1992; Chang et al. 2000), cow manure (La et al. 2003) and whey (Christensen et al. 1996). The use of these solid substrates has been particularly used for packed bed reactors since the substrates also function as a bacterial support.

In the food and beverage industry, paper industry, agriculture and households, organic waste streams are produced (De Mes et al. 2003). Currently organic waste streams are not used as electron donor for high rate sulfate reducing bioreactors for several reasons. In many cases the quantity and quality of the waste streams is not constant. In addition, such waste streams contain slowly degradable organic matter, which compromises its application (Bijmans et al. 2011).

1.3.2.2 Bioreactor configurations

Biological sulfate reducing reactors are mainly based on biomass retention inside the reactor. Numerous bioreactor systems have been used for sulfate reduction (Figure 1.2) including *i)* continuously stirred tank (CSTR) reactors (Barnes et al. 1991; White and Gadd 2000), *ii)* packed-bed (PBR) reactors (Maree and Strydom 1985; Elliott et al. 1998; El bayoumy et al. 1999; Kolmert and Johnson 2001; Jong and Parry 2003), *iii)* gas-lift (GLR) reactors (van Houten et al. 1994; Weijma et al. 2002; Esposito et al. 2006; Bijmans et al. 2009a), *iv)* up-flow anaerobic sludge blanket (UASB) reactors (Lettinga et al. 1980), *v)* membrane bioreactors (Vallero et al. 2005) and *vi)* fluidized-bed (FBR) reactors (as described in more detail below). Each of the above cited reactor types is characterized by benefits and drawbacks (Table 1.2).

Fluidized bed reactors

FBRs have been used for sulfate reduction for treatment of acidic metal containing wastewaters with sand as carrier material (Kaksonen et al. 2006; Kaksonen et al. 2003a). The FBR is based on the development of biofilm on particles that support bacterial growth, which allows the retention of biomass within the reactor and therefore to operate at short hydraulic retention time (HRT) (Shieh and Keenan 1986; Marin et al. 1999). Carrier materials used include iron chips (Somlev and Tishkov 1992), synthetic polymeric granules covered with iron dust (Somlev and Tishkov 1992) and silicate minerals (Kaksonen et al. 2003b). Fluidization of the carrier material is achieved by recycling the effluent from the top to the bottom of the reactor, so that a relative high upflow velocity is reached (Figure 1.3a). The FBR has various advantages over other high rate anaerobic reactors. There is no channeling or clogging in the reactors, biomass retention is good, biomass activity and treatment efficiency are high, and there is low risk of shock loads (Kaksonen et al. 2006).

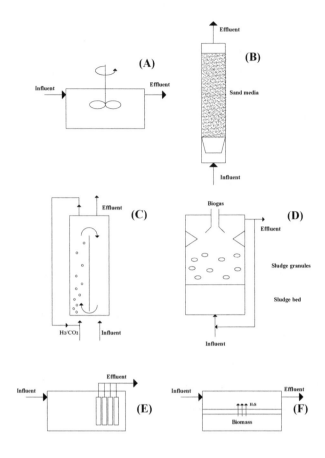

Figure 1.2 Schematic representation of the continuously stirred tank (A), packed bed (B), gas lift (C), up-flow anaerobic sludge blanket (D), immersed membrane (Geysen et al. 2004) (E) and extractive membrane (F) bioreactors (Papirio et al. 2012).

The downflow FBR, also known as inversed fluidized bed (IFB) reactor, is based on a floatable carrier material which is fluidized downward with a downflow current of liquid (Figure 1.3b). After inoculation, a biofilm develops over the support that remains at the top of the reactor. The sulfate conversion rate largely depends on the biomass adhered and formed on the carrier material. Carrier materials used for downflow FBR have a density lower than water, e.g. polyethylene (Castilla et al. 2000; Celis-García et al. 2007; Celis et al. 2009; Gallegos-Garcia et al. 2009), polystyrene spheres (Nikolov and Karamanev 1987), cork (García-Calderón et al. 1998) and extendosphere (Buffière et al. 2000; Arnaiz et al. 2003).

8

Table 1.2 Benefits and drawbacks of different bioreactor configuration operated for biological sulfate reduction (Papirio et al. 2012).

Bioreactor type	Benefits (+) / Drawbacks (-)	References
Continuously stirred tank	(+) Consistency and reliability	Barnes et al. (1991)
	(-) High SRT result in high reactor volumes	Barnes et al. (1991)
	(-) Frequent active biomass washout	Lens et al. (2003)
Packed-bed	(+) High SRT result in mower reactor volumes than CSTRs	Barnes et al. (1991)
	(+) Possibility to be operated both in up-flow and down-flow modalities	Jong and Parry (2003);Zaluski (2003)
	(-) Frequent clogging	Anderson et al. (1990)
	(-) High pressure for pumping the flow	Anderson et al. (1990)
Gas lift	(+) High mass transfer of the substrates into the bacterial agglomerates	Dijkman and Buisman (1999)
	(+) Very good mixing	Dijkman and Buisman (1999)
	(+) High rate biological kinetics if H_2 is used as electron donor	Van Houten et al. (1994; 1997)
	(-) High pressure needed for pumping the gaseous substrates inside the reactor	Lens et al. (2002)
Up-flow anaerobic sludge blanket	(+) Biomass good settling capability	Lettinga et al., (1980)
	(+) No clogging	Omil et al. (1996)
	(+) No carrier material if compared to PBR	Speece (1983)
	(-) Possibility of biomass washout	Vallero et al. (2003)
	(-) High susceptibility to the influent characteristics	Jhung and Choi, 1995
Membrane bioreactor	(+) No need for sedimentation basin	Mack et al. (2004)
	(+) High biomass retention result in high substrate degradation rates	Mack et al. (2004)
	(+) Possibility to prevent direct contact between metals and SRB in a single basin	Chuichulcherm et al. (2001); Manconi and Lens (2009)
	(-) High cost to overcome the trans-membrane pressure	Fedorovich (2000)
	(-) Periodic backwash because of the deposition of aggregates on the membrane surface	Tabak and Govind (2003b); Vallero et al. (2005)

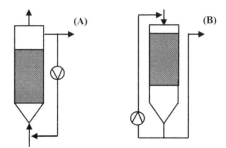

Figure 1.3 Schematic representation of an up-flow (A) and inversed (B) fluidized bed reactor (Kaksonen and Puhakka 2007).

1.3.2.3 Operational conditions of sulfate reducing bioreactors

1.3.2.3.1 pH

Low pH values increase toxicity as they promote the formation of undissociated sulfide (Hulshoff et al. 2001; Willow and Cohen 2003) and non-ionized organic acids (Kimura et al., 2006). Strong acidic wastewaters (pH 2.5-3) like AMD can be completely neutralized to pH 7.5-8.5 by the alkalinity produced by SRB during lactate or ethanol oxidation in a FBR (Kaksonen et al. 2003b). This is possible due to the alkalinity produced by the SRB and the recycle flow that dilutes influent concentrations and acidity so that SRB are active in a pH range close to that of the treated effluent. The pH increases although metal precipitation also produces acidity (Equation 2) (Kaksonen et al. 2003b).

The pH increases up to values around pH 8 due to the production of HCO_3^- by microbial metabolism (Equation 1). This has also the advantage to reduce the amount of non-ionized sulfide which is the most toxic form of sulfide (Hulshoff et al. 2001).

Even though sulfate reduction increases the pH of the effluent, this can occur under HRTs that allow dilution of the influent and adequate bicarbonate production rates. The application of sulfate reduction at low pH opens the possibility to selectively recover metals from multi metal streams by varying the pH and sulfide concentration (Huisman et al. 2006; Tabak et al. 2003).

Previous studies on sulfate reduction in the acidification stage have shown a substantial decrease in sulfate reduction efficiencies when the reactor pH became lower than 6.2 (Reis et al. 1992). More recently, Lopes et al. (2007) reported nearly complete sulfate reduction efficiencies during the acidification of sucrose at pH 6 and 5 at influent COD to SO_4^{2-} ratios of 9 and 3.5. This ratio is an important parameter affecting the competition between SRB and other anaerobic bacteria involved in wastewater treatment.

A shortcoming of most studies on sulfate reduction under acidic conditions is the short duration and the lack of pH control during the experiment. Little is known about sulfate reduction under acidic conditions in bioreactor runs under controlled pH. Sulfate reduction at pH 3.8 has been reported in a batch reactor fed with glycerol (Kimura et al., 2006) and at pH 4 in a sucrose fed upflow anaerobic sludge bed (UASB) reactor (Lopes et al. 2007) and in an membrane bioreactor (Geysen et al. 2004) with formate and hydrogen as electron donor (Bijmans et al. 2010). The achieved rates limit industrial application and SRB growth has not yet been demonstrated at these low pH values. A potential cause for the lack of growth is that the organic electron donor was partly converted to acetate, which has been shown to be toxic under acidic conditions.

1.3.2.3.2 HRT

Several studies assessed the effect of the HRT on a FBR treating a heavy metal and sulfate containing wastewater. In almost all studies, a higher HRT is applied at the beginning of the experiments to enhance the contact time between the microorganisms and the support (Celis et al. 2009). This results in a significant biomass immobilization on the carrier material. Then, in order to test the robustness of the reactors, the HRT is quickly or gradually decreased. If a significant biofilm development is attained on the carrier support, pseudo-steady states are reached in a short period of time after operational HRT changes (Celis-García et al. 2007).

Kaksonen et al. (2004) demonstrated that an acidic fed pH-FBR can be successfully operated at low HRTs (6.5 h) if the HRT is gradually decreased. Effluent dissolved zinc and iron concentrations were below 0.1 mg/L during stable performance of the reactor at an HRT of 6.5 h (Kaksonen et al. 2004). However, a sudden decrease of the HRT from 9.7 h to 7.3 h resulted in a decrease of the reactor performance since higher metal concentrations were detected in the effluent solution. In contrast, Villa-Gomez et al. (2011) did not observe a relevant variation in metal precipitation with a sudden change of HRT from 24 to 9 hours in two IFB reactor treating a wastewater containing Zn, Cu, Pb and Cd operated at different sulfide concentrations.

In terms of sulfate reduction performance, Nagpal et al. (2000) observed that a gradual decrease of the HRT from 35 h to 5 h led to a decrease of the sulfate reduction efficiency (from 90% to 65%) in the reactor at the highest influent sulfate concentration (2.5 g/L). At an HRT of 5 h the maximum sulfate reduction rate was reached (6.33 g/L · day) showing that the liquid–solid fluidized-bed process responded well to the intentional gradual HRT changes removing a higher sulfate quantity than during the previous experimental period. At low HRTs, incomplete ethanol oxidation resulted in an increase of the acetate effluent concentration, especially when the HRT was decreased below 12 h (Kaksonen et al. 2004).

1.3.2.3.3 Sulfide inhibition

The inhibitory effect of sulfide and organic acids is caused when these substances are present as unionized compounds because only neutral molecules permeate the cell membrane (Oleszkiewicz et al. 1989; Hulshoff et al. 2001; Kimura et al. 2006) Undissociated acids enter the bacterial cell, acidify the cytoplasm and lead to bacterial death at high concentrations (Kimura et al. 2006; Celis-García et al. 2007). Celis-Garcia et al. (2007) showed that, although the sulfide concentration reached a value of 1215 mg/L, both lactate consumption and sulfate reduction were not affected by high sulfide production and kept stable around 90% and 75% removal efficiency, respectively. The high recirculation rate may have contributed to the formation of a biofilm able to tolerate high total sulfide concentrations without any apparent toxic effect.

1.3.2.3.4 Inhibition by heavy metals

Heavy metals can affect the SRB metabolism by deactivating the enzymes and denaturing the proteins (Cabrera et al. 2006). According to the metal concentration, the effect can be different: *i)* inhibition of the bacterial growth, *ii)* extension of the lag phase in sulfide production, *iii)* decrease in sulfate-reducing activity and *iv)* death of bacteria (Sani et al. 2003; Cabrera et al. 2006). However, the sulfide reacts quickly with metals forming metal sulfide particles, which reduces the metal toxicity and bioavailability.

1.4 Sulfate reduction and metal precipitation process configuration
1.4.1 Single stage process

Many bioreactor configurations have been reported in the last years (section 1.3.2.4); however only some of them have been applied for sulfate reduction and metal precipitation in a single stage (Table 1.3). These include UASB reactors (Kaksonen et al. 2003b; Sierra-Alvarez et al. 2006; Steed et al. 2000), UAPBR reactors (Dvorak et al. 1992; Hammack et al. 1994; Chang et al. 2000; Jong and Parry 2003; La et al. 2003), fixed-bed reactor (Foucher et al. 2001), FBRs (Kaksonen et al. 2003b; Kaksonen et al. 2004), down-flow FBRs (Gallegos-Garcia et al. 2009), upflow packed anaerobic filter (Steed et al. 2000) and GLRs (Hammack et al. 1994).

The single stage treatment process is a low-cost solution for AMD treatment, but it may not be viable if the wastewater is very acidic or contains high concentrations of heavy metals (Hao et al. 1994). The optimum pH for SRB is usually stated as circum neutral. Studies in Table 1.3 show pH influents below 3; however this pH is increased due to the production of alkalinity during treatment by SRB which shows that the actual treatment could be performed at pH values higher than that of the waste stream. Sulfate reduction has been described under acidic conditions as low as 3.8 (Kimura et al. 2006). In a single stage under acidic conditions, however, biological sulfate reduction and metal precipitation are reported at the lowest pH of 5 in the influent and effluent (Bijmans et al. 2008; Gallegos-Garcia et al. 2009).

The interaction between the metals and the biomass is also a concern in single-stage processes due to metal toxicity. Depending on the sludge characteristics, the chemistry of the mixed liquor (e.g. sulfide concentration and pH), the dissolved metal species and their respective concentrations, SRB is capable of binding and accumulating high quantities of heavy metals (Labrenz et al. 2000). In addition, the exposure time determines the outcome of toxicity. Studies of short duration in minimal media have indicated that biofilms are highly tolerant to heavy metals, whereas long term exposures in rich media induce a complete inhibition of biomass (Teitzel and Parsek 2003).

1.4.2 Two stage and multi stage process

Sulfate reduction and metal removal can be performed in a two-stage process consisting of a biological stage, separated from the precipitation stage. The major advantage is the avoidance of biomass toxic effects due to high acidity and metal concentrations in the AMD streams (Johnson and Hallberg 2005). Tabak et al. (2003) proposed a two stage process in which the metal precipitation step was separated from the SRB bioreactor system. In the last decade, this technique has been studied with metals like Cu and Zn (Foucher et al. 2001; Al-Tarazi et al. 2005b; Gramp et al. 2006; Esposito et al. 2006). Selective precipitation of individual heavy metals can be realized by controlling the pH and pS (pS=-log(S^{2-})) (Veeken et al. 2003b; Esposito et al. 2006; König et al. 2006; Sampaio et al. 2009). This results in pure precipitates of metal sulfides that can be reused as a raw starting material in the metal industry (Grootscholten et al. 2008)

Metal recovery is another drawback of single-stage treatment process configuration since metal sulfides can precipitate in the biomass. Although the metals can be recovered from the metal sulfide-containing sludge (Tabak et al. 2003), this might imply biomass lost. In this sense, IFB reactors are the most suitable reactor configuration for metal recovery separately from the biomass.

Table 1.3 Lab-scale reactors used for simultaneous biological sulfate reduction and heavy metal precipitation.

Reactor type	HRT (h)	Substrate	T (°C)	Influent pH	COD/SO_4^{2-} ratio	COD (mg/L)	COD remov	Metal	Metal concentration	Metal removal	Reference
UAPBR	12	Acid-washed mushroom compost[e] lactate	-	4.5	1.9[a]	3733.33[ad]	-	Ni	500	> 99	Hammack and Edenborn, 1992
UAPBR	16.2	Lactate	25	~ 4.5	2	5000	-	Cu	50, 10, 20, 5	> 97.5	Jong and Parry, 2003
								Zn	50, 10, 20, 5	> 97.5	
								Ni	50, 10, 20, 5	> 97.5	
								As	50, 10, 20, 5	> 77.5	
								Fe	50, 10, 20, 5	> 82	
								Mg	50, 10, 20, 5	0	
								Al	50, 10, 20, 5	0	
UAPBR	480	Compost[b]	25	6.8	-	-	-	Fe	500	< 80	Chang et al., 2000
								Zn	100	> 99	
								Mn	50	< 1	
								Cu	50	> 99	
UAPBR	96	cow manure	20 - 23.5	2.6 - 2.9	-	-	-	Mg	122	> 99	La et al., 2003
								Fe	279	> 99	
								Zn	33	> 99	
								Mn	27	> 94	
								Al	13	> 99	
								Cu	68	> 99	
								Cd	56	> 99	
FBR	16	Lactate	35	2.5	0.66[a]	~ 1500[a]	> 98	Zn	166 - 233	> 99	Kaksonen et al., 2003
								Fe	58	> 99	
FBR	6.5	Ethanol	35	3 - 3.2	0.72[a]	1440[a]	-	Zn	200	> 98	Kaksonen et al., 2004
								Fe	100	> 98	
Down-flow FBR	24	Ethanol/ Lacate	18 - 26	5 - 6	0.8	2500	~ 50	Fe	140 - 320	> 98	Gallegos-Garcia et al., 2009
								Zn	40 - 220	> 98	
								Cd	5 - 6	> 99	
AFR	202	Acetic acid	30	7.2	0.7	3500	> 95	Fe	840	> 99	Steed et al., 2000
								Zn	650	> 99	
								Mn	275	> 97	
								Cu	127	> 99	
								Cd	2.3	> 99	
								As	2.1	> 99	
								Pb	1.5	> 99	
UASB	16	Lactate	35	3	0.66[a]	~ 1083[a]	> 98	Zn	166 - 233	> 99	Kaksonen et al., 2003
								Fe	58	> 99	
UASB	24	Ethanol	30	4.5	1.28	0.9	90.5 - 85.8	Cu	100	100	Sierra-Alvarez et al., 2006
								Zn	15	99.5	
								Ni	15	99.6	
GLR	98.4	Lactate syrup	24 - 30	2.3	-	3861.8[a]	-	Fe	178	> 99	Hammack et al., 1994
								Cu	530	> 99	
								Zn	620	> 99	
								Al	278	> 99	
								Mn	191	> 91	

UAPBR: Upflow anaerobic packed bed reactor, HRT: Hydraulic retention time, pH influent, [a]Values estimated from raw data, [b]Oak chips,spent oak chips,organic rich soil, spent mushroom, paper sludge (each substrate was used in one reactor but the results were similar), [c]Partially packed with floating plastic pall rings; [d]Calculation based on lactate; [e]Composed of straw, hay, horse and poultry manure, ground corncobs, gypsum and limestone.

1.5 Metal sulfide precipitation process

Besides the reactor configuration and the number of stages in the treatment process, metal sulfide precipitation itself is a complex process that needs to be understood for metal recovery. Precipitation occurs through several steps, namely nucleation, crystal growth, aggregation and eventually breakup (Al-Tarazi et al., 2004). Most of the fundamentals of precipitation theory are well understood and therefore, only a general summary is given in box 1.

Metal sulfide precipitates have low solubility products (Table 1.4) and, as a consequence, very small particles normally prevail as the supersaturation level cannot be controlled at low levels (Hammack et al., 1994) and because local supersaturation at the feed points due to micromixing limitation cannot be prevented (Tabak et al., 2003). As a result, metal sulfide precipitation reactions are difficult to control and a large number of small particles are formed during the process (Mokone et al., 2010).

Metal sulfide species can be present as metal-sulfide complexes, nanoclusters (in the range of 2-10 nm) or colloids (small particles in the range 1nm-1μm) that are formed in supersaturated solutions (Sukola et al. 2005). Van Hille et al. (2005) found that elevated supersaturation causes the rapid precipitation of Cu sulfides, often resulting in the formation of fines and hydrated colloidal particles. On the other hand, Lewis and van Hille (2006) found that low sulfide concentrations lead to the formation of aqueous sulfide clusters at high supersaturation points. As a consequence of its particle size, metal sulfides are difficult to recover from the water phase by means of solid–liquid separation processes (Bhattacharyya et al. 1979).

Table 1.4 Solubility product constants of metal sulfides at the standards condition (25°C, 1atm) (Sampaio et al. 2009).

Metal ion	Log K_{SP} (metal sulfide)
Hg(II)	−52.4
Ag(I)	−49.7
Cu(I)	−48.0, −48.5
Cu(II)	−35.1
Cd(II)	−27.7, −25.8
Pb(II)	−27.0, −27.5
Zn(II)	−23.8
Ni(II)	−20.7
Fe(II)	−17.3

□

Box 1. Precipitation theory

Nucleation is the step where the solute molecules dispersed in the solvent start to gather into clusters on the nanometer scale (elevating solute concentration in a small region), that becomes stable under the current operating conditions. These stable clusters constitute the nuclei. Nucleation produces only very small particles which are difficult to separate from the liquid phase, whereas crystal growth and agglomeration of crystals result in larger particles. The crystal growth is the subsequent growth of the nuclei that succeed in achieving the critical cluster size. Crystallization is the formation of a solid state of matter in which the molecules are arranged in a regular pattern (Larsen et al., 2006).

The saturation concentration, also called solubility, represents the minimum solute concentration for which crystal growth can occur. The supersaturation level is the amount by which the solute concentration exceeds the saturation concentration. Crystallization occurs only if the system is supersaturated (Larsen et al., 2006). In general, high supersaturation levels favour nucleation, thus to produce large particles the supersaturation should be minimized. The supersaturation level σ for metal sulfides can be expressed in terms of the solubility product (Veeken et al., 2003b):

$$\sigma = \sqrt{\frac{\left(Me^{2+}\right)\left(S^{2-}\right)}{K_{sp}}}$$

σ= the supersaturation level , (Me^{2+})= metal activity (mol/l), (S^{2-})= sulfide activity (mol/l).

Nucleation and growth continue to occur simultaneously while the supersaturation exists (Figure). Supersaturation is the driving force of the crystallization. Depending on the conditions, either nucleation or crystal growth may be predominant over the other, and as a result, crystals with different sizes and shapes are obtained.

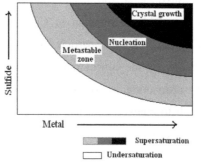

Simplified phase diagram of metal sulfide precipitation.

During precipitation, a dispersed solid-liquid system is formed. The stability of this dispersed system is determined by the extent to which the original degree of dispersion is maintained over a period of time. The initial dispersion may change as a result of formation of larger entities by particle aggregation. This process is called coagulation for small particles and agglomeration for larger particles. Aggregate formation depends on the transport of the primary particles due to Brownian motion, fluid motion or sedimentation, frequency of collision between moving particles and the nature of interactions that take place when these particles collide (Söhnel and Garside, 1992). The latter may be attractive or repulsive depending on the surface characteristics of the particles and the chemistry of the solution.

Many studies on metal sulfide precipitation have focused on studying the way to reduce the high level of supersaturation on the metal sulfide precipitation in order to increase the size of the precipitates for better solid-liquid separation. Seckler (1996) suggested the use of a seeded fluidized bed reactor with multiple reagent feed points. This approach was used by Van Hille et al. (2005), who studied the influence of the sulfide to Cu molar ratio, recycle flow rate, inlet Cu flow rate and the inlet copper concentration on the Cu conversion and removal efficiency. They found that the sulfide to Cu molar ratio and the bisulfide ion formation were the most important factors determining local supersaturation. Al-Tarazi (2004) used gaseous H_2S as precipitating reagent to reduce the high level of supersaturation in the precipitation of Cu and Zn sulfide in a bubble column and concluded that the morphology of the metal sulfide precipitate produced using gaseous hydrogen sulfide was more favorable than that of the precipitate produced using aqueous sulfide source.

Some studies have shown the influences of geometry and operating conditions of the precipitator reactor on the crystallization process. For instance, Al-Tarazi et al. (2005a) studied the effects of the reactor configuration on three different types of reactors for the precipitation of Zn and Cu: laminar jet, bubble column and a Mixed Solution Mixed Product Removal (MSMPR) reactor, studying the effects of mass transfer and process conditions on the morphology of the produced crystals. They found that the largest crystals of metal sulfides were obtained at high supersaturation concentrations, moderate stirrer speeds, short residence times, a pH value of around 5 and high Cu^{2+} to S^{2-} ratios. Sampaio et al. (2009) observed that the particle size of CuS increased if allowed to settle (from 36 to 180 µm), whereas upon vigorously stirred, the particles decreased to below 3 µm in experiments done in a continuously stirred tank reactor.

In addition to the influence of geometry and operating conditions, foreign particles also determine particle size of the metal sulfides by affecting the relative rates of nucleation and crystal growth. According to experiments carried out with foreign particle seeds added to the reactants, the nucleation is reduced to approximately 10% of the value for homogeneous primary nucleation (Schubert and Mersmann 1996). Furthermore, the rate of heterogeneous primary nucleation is proportional to the specific surface area of the foreign particles present in the solution (Mersmann 1999). Gramp et al. (2006) showed the differences between biogenic and abiotic sulfide used to precipitate copper in cultures of SRB and Na_2S solutions. They found that bacterial cells alter crystal formation by inhibiting particle nucleation and as a consequence the chemically produced covellite (CuS) should be more resistant to biogeochemical oxidation as compared to poorly crystalline biogenic Cu-sulfide. Contrary to the previous authors, Bijmans et al. (2009) suggested that the biomass functioned as nucleation seeds, enhancing crystal growth, reporting NiS precipitates formed with biogenic sulfide ranging from 13-73 μm.

Foreign particles can also promote or impede aggregation of the metal sulfides that affects particle size. Peters et al. (1984) found that, for ZnS precipitated at pH 8, the complexing agents (ammonia, EDTA and 18-Crown-6 ether) reduced the nucleation rate, but promoted the aggregation rate. Esposito et al. (2006) found that the use of biogenic sulfide instead of Na_2S as sulfide source for ZnS precipitation decreased the efficiency of the precipitation process both in terms of zinc effluent concentration and particle size of the precipitates due to the presence of substances commonly present in bioreactors such as ammonium and phosphate.

1.5.2.1 Factors affecting metal sulfide precipitation

1.5.2.1.1 pH

The pH is important because it influences both the solubility of sulfides, and the kinetics of the precipitation processes for selective metal recovery. As shown in Equation 4 and 5, aqueous sulfide speciation and thus the aqueous concentration of the reactive sulfide species (i.e. HS$^-$ and S^{2-}) is dependent on the pH of the solution. Figure 1.4 shows the pH-dependent aqueous sulfide speciation (MEDUSA, software implemented by The Royal Institute of Technology, Sweden). It can be observed that for an increase of one pH unit, the S^{2-} concentration increases by two orders of magnitude. Hammack et al. (1994) investigated the effect of pH on Cu sulfide precipitation using biologically produced gaseous hydrogen sulfide. The rate of gaseous H$_2$S dissolution was found to be low under acidic conditions and this led to a decrease in the total dissolved sulfide concentration required for metal precipitation.

$$H_2S_{(aq)} \leftrightarrow HS^-_{(aq)} + H^+_{(aq)} \qquad pK_1 = 7 \qquad (3)$$

$$HS^-_{(aq)} + H^+_{(aq)} \leftrightarrow S^{2-}_{(aq)} + 2H^+_{(aq)} \qquad pK_2 = 13.9 \qquad (4)$$

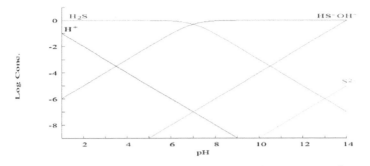

Figure 1.4 Aqueous hydrogen sulfide speciation as a function of pH.

Solubility diagrams are used in order to predict the precipitation efficiency. Figure 1.5 shows the solubility of several metal sulfides in distilled water at various pH values. The pH-dependent solubility of different metal sulfides has been successfully used to selectively recover metal ions from complex and simple mixed metal systems (Tokuda et al. 2008; Bijmans et al. 2009b; Sahinkaya et al. 2009; Tabak et al. 2003). Foucher et al. (2001) found that Cu and Zn could be selectively recovered at pH 2.8 and pH 3.5, respectively. Other metals such as Ni and Fe could only be removed (not recovered) at pH 6 by sulfide precipitation. Similar results were found by Sampaio et al. (2008) who selectively precipitated Cu (covellite) at pH 2 and 3 and ZnS (sphalerite) at pH 3 and 4. Bijmans et al. (2009) demonstrated that Ni can be selectively recovered from a Ni-Fe solution at pH 5 using a single stage bioreactor operating at low pH. The results also suggested that the pH should be lower than 4.8 for complete Ni-Fe separation.

1.5.2.1.2 Solubility product

The solubility product determines whether a metal sulfide will stay dissolved or will precipitate. For a solid precipitate of metal sulfide M_xS_y (s), the following general solubility expression can be written:

$$M_xS_y \text{ (s)} \longleftrightarrow x\ M^{2+} + y\ S^{2-} \tag{5}$$

The solubility product (K_{sp}) of the metal sulfide is defined as:

$$K_{sp} = [M^{2+}]^x * [S^{2-}]^y \tag{6}$$

Where *Ksp* is in mol^2/l^2 when x=y=1, $[M^{2+}]$ is the equilibrium activity of metal ion M^{2+} (mol/l) and $[S^{2-}]$ is the equilibrium activity of S^{2-} (mol/l).

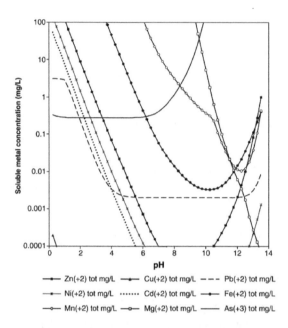

Figure 1.5 pH dependence of metal sulfide solubility (Lewis 2010).

Sulfide gives the possibility of selective precipitation due to the different solubility products of the different metal sulfides (Table 1.3). Having the solubility product defined as (6), it means that different sulfide concentrations (S^{2-} potentials) are required to precipitate different metals. Several authors have found that the equilibrium in the metal sulfide reactions are ruled by the sulfide concentrations by means solubility product (Bryson and Bijsterveld 1991; Mishra and Das 1992; Lewis and Swartbooi 2006; Veeken et al. 2003b). Veeken et al. (2003) showed that at pH 6 different metals (Cd, Cu, Ni, Pb and Zn) had different sulfide curves expressed in the logarithm of the S⁻ species (pS) curves when titrated and precipitated with sulfide.

1.5.2.1.3 Competing metal removal mechanisms

Several mechanisms, such as sorption onto the biomass (Neculita et al. 2007), complexation and precipitation with other compounds typically present in the effluent of a sulfate reducing bioreactor such as extracellular polymeric substances (Beech and

Cheung, 1995), complexing agents (organic compounds) (Peters et al., 1984), phosphate and micronutrients from the biological process (Esposito et al., 2006; König et al., 2006) compete with the metal sulfide precipitation process, thus jeopardizing the success of the metal sulfide recovery.

Sorption mechanisms in the biomass are explained as either through an ion exchange mechanism on the surface of the biosorbent (biofilm) or surface precipitation of metal hydroxide/sulfide species (van Hullebusch et al. 2003). A comparison between batch growth tests (bioprecipitation) and dead biomass batch tests (biosorption) with SRB by Pagnanelli et al. (2010) showed that Cd was mainly removed by a biosorption mechanism (77%) due to metabolism-independent binding properties of the SRB cell wall surface. A sorption mechanism is described by Jong and Parry (2003) occurring through the formation of strong, inner–sphere complexes involving surface hydroxyl groups on the bacterially produced metal sulfides (BPMS). That is, the BPMS surfaces are hydrated in aqueous solution by water, producing surface hydroxyl (-OH) groups. This is followed by a ligand exchange reaction between the surface hydroxyl groups and metal–adsorbing ions, leading to the formation of inner-sphere metal complexes.

Metal removal mechanism involving precipitation with compounds typically present in the effluent of a sulfate reducing bioreactor are due to the presence of hydroxides, carbonates and phosphates, contained in the synthetic wastewater. Hydroxide precipitation of metals has been reported in passive sulfate reducing reactors operating at a pH around 7 during the first stages of operation, when sulfate reduction is not yet well established and hence, low sulfide production occurs (Samaranayake et al. 2002; Neculita et al. 2008). Several studies have shown the potential of phosphates for the immobilization of divalent heavy metals like Zn, Pb, Cu, and Cd from wastewaters, solid wastes and contaminated soils (Ruby et al. 1994; Eighmy et al. 1997; Geysen et al. 2004), surprisely, no evidence of phosphate precipitation is reported in sulfate reducing reactors. Bartacek et al. (2008) showed that cobalt depletion might be due to phosphates and carbonates precipitation in a study on the influence of cobalt speciation on the toxicity of cobalt to methylotrophic methanogenesis in anaerobic granular sludge evidencing the possibility of this removal mechanism in sulfate reducing reactors.

1.6 Optimization of the sulfate reducing process for metal precipitation

1.6.1 Modeling

Mathematical modeling is a powerful tool for process analysis and design. It also forms the basis for monitoring and control schemes of bioreactors. Sulfate reduction modeling has been well developed in several reactor configurations. These studies have been made for the modeling of sulfide toxicity (Kaksonen et al., 2004), competition of SRB with methanogenesis (Kalyuzhnyi and Federovich, 1997; Chou et al., 2008), kinetic parameters of sulfate reduction (Nielsen, 1987, Tabak et al., 2003), biofilm growth (Nielsen, 1987), and scale-up design (Tabak et al., 2003).

The metal precipitation process has been modeled separately from the sulfate reducing process in CSTR's to predict the effects of organic substances (Peters et al., 1984, König et al., 2006), sulfide concentration (König et al., 2006), pH variation (Luptakova and Kusnierova, 2005; König et al., 2006; Sampaio et al., 2008) and metal concentration (König et al., 2006) on the metal precipitation, to obtain crystallization kinetics (Peters et al., 1984; Al-Tarazi et al., 2004), and to design an adequate control strategy to estimate the effluent metal concentration (Sampaio et al., 2008).

Mathematical modeling has facilitated the design of controllers for metal precipitation. These tools can predict optimal conditions like pH, temperature, and sulfide concentration, thus reducing costs on reactor construction and operation. However, these experiments have been carried out separated from the biological sulfate reduction process and when metal recovery is combined with biological sulfate reduction, the process is restricted with respect to the operational pH, temperature, and sulfide concentration (Widdel, 1988 #198). For the modeling of the process it is important to consider that if the sulfate reduction and metal precipitation take place in a single stage reactor, metals can also interact with the bacteria themselves (Jong and Parry, 2004; Gramp et al., 2006) and their excreted components (Beech and Cheung, 1995). Moreover, in general, modeling tools have only studied the sulfide precipitation of Zn and Cu. Therefore, it is necessary to make studies with other metals as well, so that the models can be adjusted for these other metals.

1.6.2 Sulfide control

Sulfide control production in sulfate reducing bioreactors is highly relevant to avoid the unnecessary production of sulfide that increases operational costs and may impose a sulfide removal post-treatment step. The principles of monitoring and control in bioreactors are reviewed in Dunn et al. (2005) and Pind et al. (2003). A general block diagram representing a control configuration is given in figure 1.6. Biological process control has been studied for several desirable end products such as ethanol, penicillin, yeast, wastewater treatment and diverse fermentations (Dunn et al. 2005). Set-point control in these processes is typically based on the manipulation of temperature, pH, substrate or dissolved oxygen. The key issues in process control are the use of a suitable sensor and selecting an appropriate control strategy and control parameters. In the past years a method using a sulfide-selective electrode (pS electrode) to control the sulfide addition has been studied by several authors (Grootscholten et al. 2008; König et al. 2006; Veeken et al. 2003a). These studies showed that the stoichiometric addition of sulfide to a solution containing heavy metals can be achieved using the pS electrode combined with a pH electrode resulting in very low concentrations of both metals and sulfide. These experiments have been carried out separate from the biological sulfate reduction process. Studies on the control of the sulfide production in bioreactors are lacking in the literature.

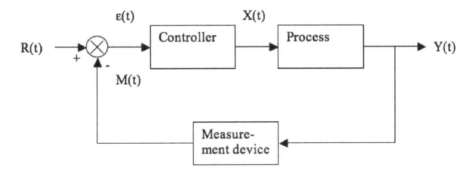

Figure 1.6 Block diagram representing a control system. (t) = "set point" is a synonym for the desired value of the controlled variableε, (t) = process error or comparator, $X(t)$ = process variable, $M(t)$ = measurement variable, $Y(t)$ = output variable

Even though large progress has been made in modeling and control of anaerobic systems, there is still insufficient knowledge in control of sulfate reducing bioreactors. Some authors have designed algorithms for control of the pH, and concentration of organic compounds, by controlling the biodegradation of organic compounds (Aguilar et al. 2001; Mailleret et al. 2004; Ryhiner et al. 1993). Ryhiner et al. (1993) used the flow rate as the manipulated variable to control the pH, dissolved hydrogen and organic acids concentration in simulation and experiments. The experimental results were found in close agreement with predetermined simulated values. Mailleret et al. (2004) developed a nonlinear controller for its application in anaerobic wastewater treatment plants. The controller was shown in simulations, while experiments were missing. Although some authors have developed mathematical models for sulfate reduction in anaerobic digestion (Kalyuzhnyi and Fedorovich 1998; Oyekola et al. 2012; Gupta et al. 1994), they have focused on the population dynamics and substrate competition among microbial species while accounting for sulfide production.

The selection of the most adequate control strategy largely depends on the process characteristics. A proportional-integral-derivative (PID) controller has been widely used in bioreactors partly due to the derivative time parameter used to overcome the lag presence caused by the time required for substrate degradation before being used or transformed to the desired product (Dunn et al. 2003; Dunn et al. 2005; Jagadeesh and Sudhaker 2010). This controller has three adjustable parameters K_c controller gain, τ_i integral time and τ_d derivative time. These parameters can be obtained by using different tuning methodologies.

1.6.3 Metal speciation techniques in sulfate reducing reactors

Since metal sulfides crystallize poorly or are found immerse on a solution matrix when biogenic sulfide is used, their analysis requires species-specific analytical methods (Neculita et al. 2007). In some studies, the number of techniques for collecting mineralogical data has been limited by the poor crystallinity of the precipitates and the relatively low concentrations of metal sulfides (Neculita et al. 2007). Since many samples from environmental matrices are either non-crystalline or in a solution matrix, XRD may not be the best method to investigate the structure of environmental samples. Another disadvantage of XRD is that many samples are not concentrated enough to be analyzed using XRD (Parsons et al. 2002).

XAS, which consists of two complementary techniques, the X-ray absorption near edge spectroscopy (XANES) and extended X-ray absorption fine structure (EXAFS) is a technique that has proven to be a powerful technique in environmental sciences (Parsons et al. 2002). XAS provides a clear picture of the local coordination environment of an element in both simple and complex mixtures of compounds (Parsons et al. 2002). XANES and EXAFS analysis have been applied in samples obtained from bioreactors (Lenz et al. 2011; Villa-Gomez et al. 2012; Prange and Modrow 2002).

Another powerful technique to predict speciation is the use of geochemical models of water chemistry data in bioreactors such as WATEQ4F and VMINTEQ. These models give useful insights into the chemical equilibrium reactions potentially controlling the concentrations of dissolved metals (Neculita et al., 2007).

1.7 Research needs

Different reactor configurations have been tested for sulfate reduction and metal precipitation (section 1.4). However, metal recovery cannot always be achieved in these reactors, since metals precipitate partly in the biomass, which hampers their recovery. A suitable reactor configuration for metal sulfate reduction and metal removal in a single unit is the inversed fluidized bed (IFB) reactor (section 1.3.2.2).

Metal sulfide precipitation and recovery not only depend on the reactor configuration but also on the nature and morphology of the metal precipitates for metal sulfide settling and removal. Applications of metal precipitation with sulfide are still limited by the challenging liquid-solid phase separation due to the predominance of nucleation over crystallization (Mokone et al. 2010). This leads to the formation of small, poorly settling particles or even sulfide clusters at high supersaturation conditions (Lewis and Swartbooi 2006). Therefore, it is important to study the bioreactor conditions that affect the size and structure of the metal precipitates, for example, the substances commonly present in the bioreactor mixed liquor such as dissolved organic matter (van Hullebusch et al. 2003) and macronutrients (Esposito et al. 2006; Sampaio et al. 2009; Bijmans et al. 2009a) and the hydrodynamic conditions(Al-Tarazi et al. 2005a; Al-Tarazi et al. 2005b).

Furthermore, steering sulfide production in bioreactors is highly relevant to avoid the over production of H_2S that increases operational costs and may impose a sulfide removal post-treatment step. The development of a control strategy based on the sulfide concentration as control input parameter is necessary for the metal recovery further optimization.

Also the low pH of the metal-containing wastewaters such as AMD limits the metal recovery in single stage applications due to SRB inhibition, lack of growth and low sulfate reduction rates (Bijmans et al. 2010). A clear understanding about how operational parameters can help to achieve sulfate reduction at low pH has thus far not been analyzed.

1.8 Scope and organization of this thesis

The main objective of this thesis was to elucidate the factors affecting simultaneous sulfate reduction and selective recovery of heavy metals in an IFB reactor in order to optimize metal recovery from metal containing wastewaters such as AMD.

The present dissertation comprises eight chapters. This chapter gave a literature review about the current knowledge of both sulfate reduction and metal precipitation processes. Based on that knowledge, the possibilities and limitations for metal recovery in an inverse fluidized bed reactor are addressed.

Supersaturation, which depends on the stoichiometry of the reactants, is a key factor in understanding metal sulfide precipitation inside bioreactors. In Chapter 2, the effect of the sulfide concentration on the location of metal precipitates within the IFB reactor was evaluated.

Chapter 3 explores the influence of the of sulfide and macronutrient concentration commonly present in mineral media and wastewaters on the metal removal kinetics and mechanisms. Chapter 4 analyzes the differences in the metal precipitation characteristics as a function of the decrease in the HRT in IFB reactors. The effect of the reactor operational changes on the SRB population in the biofilm and suspended biomass was also described.

Chapter 5 investigates key components contained in the biogenic sulfide that affect the nature and morphology of the metal precipitates (Cu and Zn). Metal precipitation experiments using biogenic sulfide were performed for particle size, settling rate and dewaterability analysis of the metal precipitates under different pH values.

Chapter 6 describes the individual and combined effect of COD in the influent, pH, metal to sulfide ratio and HRT on sulfide production, on the sulfate reduction efficiency at low pH. Different reactor runs at different conditions were analyzed using factorial design analysis to determine how these variables affect low pH sulfate reduction.

Chapter 7 evaluates input strategies to control the sulfide concentration in bioreactors using a pS electrode. The experiments were designed to create a response to the signal of the pS electrode in order to determine optimal control parameters for the PID control. Chapter 8 concludes this thesis, discussing the implications of the obtained results for metal recovery in inversed fluidized bed reactors and giving suggestions for future research.

REFERENCES

Aguilar R, González J, Barrón M, Martın-Guerra R, Maya-Yescas R (2001) Robust PI2 controller for continuous bioreactors. Process Biochemistry 36 (10):1007-1013. doi:10.1016/s0032-9592(01)00133-9

Al-Tarazi M, Heesink ABM, Azzam MOJ, Yahya SA, Versteeg GF (2004) Crystallization kinetics of ZnS precipitation; an experimental study using the mixed-suspension-mixed-product-removal (MSMPR) method. Crystal Research and Technology 39 (8):675-685. doi:10.1002/crat.200310238

Al-Tarazi M, Heesink ABM, Versteeg GF (2005a) Effects of reactor type and mass transfer on the morphology of CuS and ZnS crystals. Crystal Research and Technology 40 (8):735-740

Al-Tarazi M, Heesink ABM, Versteeg GF, Azzam MOJ, Azzam K (2005b) Precipitation of CuS and ZnS in a bubble column reactor. AIChE Journal 51 (1):235-246

Arnaiz C, Buffiere P, Elmaleh S, Lebrato J, Moletta R (2003) Anaerobic Digestion of Dairy Wastewater By Inverse Fluidization: the Inverse Fluidized Bed and the Inverse Turbulent Bed Reactors. Environmental Technology 24 (11):1431-1443

Aziz HA, Adlan MN, Ariffin KS (2008) Heavy metals (Cd, Pb, Zn, Ni, Cu and Cr(III)) removal from water in Malaysia: Post treatment by high quality limestone. Bioresource Technology 99 (6):1578-1583. doi:10.1016/j.biortech.2007.04.007

Barnes LJ, Janssen FJ, Sherren J, Versteegh JH, Koch RO, Scheeren PJH (1991) New process for the microbial removal of sulphate and heavy metals from contaminated waters extracted by a geohydrological control system. Chemical Engineering Research and Design 69 (3):184-186

Bartacek J, Fermoso F, Baldó-Urrutia A, van Hullebusch E, Lens P (2008) Cobalt toxicity in anaerobic granular sludge: influence of chemical speciation. Journal of Industrial Microbiology & Biotechnology 35 (11):1465-1474. doi:10.1007/s10295-008-0448-0

Bechard G, Yamazaki H, Gould WD, Bedard P (1994) Use of cellulosic substrates for the microbial treatment of acid mine drainage. Journal of Environmental Quality 23:111-116

Bhattacharyya D, Jumawan AB, Grieves RB (1979) Separation of Toxic Heavy Metals by Sulfide Precipitation. Separation Science and Technology 14 (5):441-452. doi:10.1080/01496397908058096

Bijmans MFM, Buisman CJN, Meulepas RJW, Lens PNL (2011) 6.34 - Sulfate Reduction for Inorganic Waste and Process Water Treatment. In: Editor-in-Chief: Murray M-Y (ed) Comprehensive Biotechnology (Second Edition). Academic Press, Burlington, pp 435-446

Bijmans MFM, de Vries E, Yang C-H, N. Buisman CJ, Lens PNL, Dopson M (2010) Sulfate reduction at pH 4.0 for treatment of process and wastewaters. Biotechnology Progress 26 (4):1029-1037. doi:10.1002/btpr.400

Bijmans MFM, Dopson M, Ennin F, Lens PNL, Buisman CJN (2008) Effect of sulfide removal on sulfate reduction at pH 5 in a hydrogen fed gas-lift bioreactor. accepted by journal of microbiology and biotechnology

Bijmans MFM, van Helvoort P-J, Buisman CJN, Lens PNL (2009a) Effect of the sulfide concentration on zinc bio-precipitation in a single stage sulfidogenic bioreactor at pH 5.5. Separation and Purification Technology 69 (3):243-248

Bijmans MFM, van Helvoort P-J, Dar SA, Dopson M, Lens PNL, Buisman CJN (2009b) Selective recovery of nickel over iron from a nickel-iron solution using microbial sulfate reduction in a gas-lift bioreactor. Water Research 43 (3):853-861

Brooks CS (1991) Metal recovery from industrial wastes. Lewis Publishers Inc., Chelsea, MI, USA

Brüser T, Lens PNL, Truper H (2000) The biological sulfur cycle In: Lens PNL, Hulshoff PL (eds) Environmental technologies to treat sulfur pollution: principles and engineering. IWA Publishing, London, pp 47-85

Bryson AW, Bijsterveld CH (1991) Kinetics of the precipitation of manganese and cobalt sulphides in the purification of a manganese sulphate electrolyte. Hydrometallurgy 27 (1):75-84

Buffière P, Bergeon J-P, Moletta R (2000) The inverse turbulent bed: a novel bioreactor for anaerobic treatment. Water Research 34 (2):673-677. doi:10.1016/s0043-1354(99)00166-9

Cabrera G, Pérez R, Gómez JM, Ábalos A, Cantero D (2006) Toxic effects of dissolved heavy metals on Desulfovibrio vulgaris and Desulfovibrio sp. strains. Journal of Hazardous Materials 135 (1–3):40-46. doi:10.1016/j.jhazmat.2005.11.058

Castilla P, M. Meraz, Monroy O, Noyola A (2000) Anaerobic Treatment of Low Concentration Waste Water in an Inverse Fluidiized Bed Reactor. Water Science & Technology 41:245-251

Celis-García LB, Razo-Flores E, Monroy O (2007) Performance of a down-flow fluidized bed reactor under sulfate reduction conditions using volatile fatty acids as electron donors. Biotechnology and Bioengineering 97 (4):771-779

Celis L, Villa-Gómez D, Alpuche-Solís A, Ortega-Morales B, Razo-Flores E (2009) Characterization of sulfate-reducing bacteria dominated surface communities during start-up of a down-flow fluidized bed reactor. Journal of Industrial Microbiology and Biotechnology 36 (1):111-121

Chang IS, Shin PK, Kim BH (2000) Biological treatment of acid mine drainage under sulphate-reducing conditions with solid waste materials as substrate. Water Research 34 (4):1269-1277

Christensen B, Laake M, Lien T (1996) Treatment of acid mine water by sulfate-reducing bacteria; results from a bench scale experiment. Water Research 30 (7):1617-1624

De Mes TZD, Stams AJM, Reith JH, Zeeman G (2003) Methane production by anaerobic digestion of wastewater ans solid waste. In: Reith JH, Wijffels RH, Barten H (eds) Bio-methane & bio-hydrogen. Dutch Biological Hydrogen Foundation, Petten, pp 58-102

Dijkman H, Buisman CJN, Bayer HG Biotechnology in the mining and metallurgical industries: Cost savings through selective precipitation of metal sulfides, . In: Eds: S.K. Young DBD, R.P. Hackl, D.G. Dixon (ed) In Proc. of the Copper 99 – Cobre 99 International Conference, October 10-13, 1999, Vol. IV: Hydrometallurgy of Copper Phoenix, Arizona, USA, 1999. The Minerals, Metals & Materials Society, Warrendale, PA, USA, pp 113-126

Djedidi Z, Khaled JB, Cheikh RB, Blais J-F, Mercier G, Tyagi RD (2009) Comparative study of dewatering characteristics of metal precipitates generated during treatment of monometallic solutions. Hydrometallurgy 95 (1-2):61-69. doi:10.1016/j.hydromet.2008.04.014

Dries J, De Smul A, Goethals L, Grootaerd H, Verstraete W (1998) High rate biological treatment of sulfate-rich wastewater in an acetate-fed EGSB reactor. Biodegradation 9 (2):103-111. doi:10.1023/a:1008334219332

Dunn IJ, Heinzle E, Ingham J, Prenosil JE (2003) Biological Reaction Engineering: Dynamic Modelling Fundamentals with Simulation Examples second edn. Wiley-VCH Verlag GmbH &Co. KgaA, Weinheim,

Dunn IJ, Heinzle E, Ingham J, Přenosil JE (2005) Automatic Bioprocess Control Fundamentals. In: Biological Reaction Engineering. Wiley-VCH Verlag GmbH & Co. KGaA, pp 161-179. doi:10.1002/3527603050.ch7

Dvorak DH, Hedin RS, Edenborn HM, McIntire PE (1992) Treatment of metal-contaminated water using bacterial sulfate reduction: Results from pilot-scale reactors. Biotechnology and Bioengineering 40 (5):609-616

Edenborn HM (2004) Use of poly(lactic acid) amendments to promote the bacterial fixation of metals in zinc smelter tailings. Bioresource Technology 92 (2):111-119. doi:10.1016/j.biortech.2003.09.004

Eighmy TT, Crannell BS, Butler LG, Cartledge FK, Emery EF, Oblas D, Krzanowski JE, Eusden JD, Shaw EL, Francis CA (1997) Heavy Metal Stabilization in Municipal Solid Waste Combustion Dry Scrubber Residue Using Soluble Phosphate. Environmental Science & Technology 31 (11):3330-3338. doi:10.1021/es970407c

El bayoumy MA, Bewtra JK, Ali HI, Biswas N (1999) Sulfide Production by Sulfate Reducing Bacteria with Lactate as Feed in an Upflow Anaerobic Fixed Film Reactor. Water, Air, & Soil Pollution 112 (1):67-84. doi:10.1023/a:1005016406707

Elliott P, Ragusa S, Catcheside D (1998) Growth of sulfate-reducing bacteria under acidic conditions in an upflow anaerobic bioreactor as a treatment system for acid mine drainage. Water Research 32 (12):3724-3730. doi:10.1016/s0043-1354(98)00144-4

Esposito G, Veeken A, Weijma J, Lens PNL (2006) Use of biogenic sulfide for ZnS precipitation. Separation and Purification Technology 51 (1):31-39

Foucher S, Battaglia-Brunet F, Ignatiadis I, Morin D (2001) Treatment by sulfate-reducing bacteria of Chessy acid-mine drainage and metals recovery. Chemical Engineering Science 56 (4):1639-1645. doi:10.1016/s0009-2509(00)00392-4

Fu F, Wang Q (2011) Removal of heavy metal ions from wastewaters: A review. Journal of Environmental Management 92 (3):407-418. doi:10.1016/j.jenvman.2010.11.011

Gallegos-Garcia M, Celis LB, Rangel-Méndez R, Razo-Flores E (2009) Precipitation and recovery of metal sulfides from metal containing acidic wastewater in a sulfidogenic down-flow fluidized bed reactor. Biotechnology and Bioengineering 102 (1):91-99

GAO (1996) Federal Land Management: Information on efforts to inventory abandoned hard rock mines. Draft report to the ranking minority member, Committee on Resources, House of Representatives, GAO/ RCED-96-30, United States General Accounting Office, Washington DC, USA

García-Calderón D, Buffière P, Moletta R, Elmaleh S (1998) Influence of biomass accumulation on bed expansion characteristics of a down-flow anaerobic fluidized-bed reactor. Biotechnology and Bioengineering 57 (2):136-144. doi:10.1002/(sici)1097-0290(19980120)57:2<136::aid-bit2>3.0.co;2-o

Gazea B, Adam K, Kontopoulos A (1996) A review of passive systems for the treatment of acid mine drainage. Minerals Engineering 9 (1):23-42. doi:10.1016/0892-6875(95)00129-8

Geysen D, Imbrechts K, Vandecasteele C, Jaspers M, Wauters G (2004) Immobilization of lead and zinc in scrubber residues from MSW combustion using soluble phosphates. Waste Management 24 (5):471-481

Gibert O, de Pablo J, Cortina JL, Ayora C (2002) Treatment of acid mine drainage by sulphate-reducing bacteria using permeable reactive barriers: A review from laboratory to full-scale experiments. Reviews in Environmental Science and Biotechnology 1 (4):327-333. doi:10.1023/a:1023227616422

Gibert O, de Pablo J, Luis Cortina J, Ayora C (2004) Chemical characterisation of natural organic substrates for biological mitigation of acid mine drainage. Water Research 38 (19):4186-4196. doi:10.1016/j.watres.2004.06.023

Gramp JP, Sasaki K, Bigham JM, Karnachuk OV, Tuovinen OH (2006) Formation of Covellite (CuS) Under Biological Sulfate-Reducing Conditions. Geomicrobiology Journal 23 (8):613-619. doi:10.1080/01490450600964383

Grootscholten T, Keesman KJ, Lens P (2008) Modelling and on-line estimation of zinc sulphide precipitation in a continuously stirred tank reactor. Separation and Purification Technology 63 (3):654-660

Gupta A, Flora JRV, Sayles GD, Suidan MT (1994) Methanogenesis and sulfate reduction in chemostats—II. Model development and verification. Water Research 28 (4):795-803. doi:10.1016/0043-1354(94)90086-8

Hammack RW, Edenborn HM, Dvorak DH (1994) Treatment of water from an open-pit copper mine using biogenic sulfide and limestone: A feasibility study. Water Research 28 (11):2321-2329

Hao OJ, Huang L, Chen JM, Buglass RL (1994) Effects of metal additions on sulfate reduction activity in wastewaters. Toxicological & Environmental Chemistry 46 (4):197-212. doi:10.1080/02772249409358113

Harries J (1997) Acid mine drainage in Australia: its extent and potential future liability. Supervising Scientist Report 125 Supervising Scientist, Canberra

Huisman JL, Schouten G, Schultz C (2006) Biologically produced sulphide for purification of process streams, effluent treatment and recovery of metals in the metal and mining industry. Hydrometallurgy 83 (1-4):106-113

Hulshoff LWP, Lens PNL, Weijma J, Stams AJM (2001) New developments in reactor and process technology for sulfate reduction. Water Science and Technology 44 (8):67-76

J.W.H S, Elferink O, Visser A, Hulshoff Pol LW, Stams AJM (1994) Sulfate reduction in methanogenic bioreactors. FEMS Microbiology Reviews 15 (2-3):119-136. doi:10.1111/j.1574-6976.1994.tb00130.x

Jagadeesh CAP, Sudhaker RDS (2010) Modeling, Simulation and Control of Bioreactors Process Parameters - Remote Experimentation Approach. International Journal of Computer Applications 1 (10):81-88

Johnson DB, Hallberg KB (2005) Biogeochemistry of the compost bioreactor components of a composite acid mine drainage passive remediation system. Science of The Total Environment 338 (1–2):81-93. doi:10.1016/j.scitotenv.2004.09.008

Jong T, Parry DL (2003) Removal of sulfate and heavy metals by sulfate reducing bacteria in short-term bench scale upflow anaerobic packed bed reactor runs. Water Research 37 (14):3379-3389

Kaksonen AH, Franzmann PD, Puhakka JA (2004) Effects of hydraulic retention time and sulfide toxicity on ethanol and acetate oxidation in sulfate-reducing metal-precipitating fluidized-bed reactor. Biotechnology and Bioengineering 86 (3):332-343

Kaksonen AH, Plumb JJ, Robertson WJ, Riekkola-Vanhanen M, Franzmann PD, Puhakka JA (2006) The performance, kinetics and microbiology of sulfidogenic fluidized-bed treatment of acidic metal- and sulfate-containing wastewater. Hydrometallurgy 83 (1-4):204-213

Kaksonen AH, Puhakka JA (2007) Sulfate Reduction Based Bioprocesses for the Treatment of Acid Mine Drainage and the Recovery of Metals. Engineering in Life Sciences 7 (6):541-564

Kaksonen AH, Riekkola-Vanhanen M-L, Puhakka JA (2003a) Optimization of metal sulphide precipitation in fluidized-bed treatment of acidic wastewater. Water Res 37 (2):255-266

Kaksonen AH, Riekkola-Vanhanen ML, Puhakka JA (2003b) Optimization of metal sulphide precipitation in fluidized-bed treatment of acidic wastewater. Water Research 37 (2):255-266

Kalyuzhnyi SV, Fedorovich VV (1998) Mathematical modelling of competition between sulphate reduction and methanogenesis in anaerobic reactors. Bioresource Technology 65 (3):227-242. doi:10.1016/s0960-8524(98)00019-4

Kimura S, Hallberg K, Johnson D (2006) Sulfidogenesis in Low pH (3.8–4.2) Media by a Mixed Population of Acidophilic Bacteria. Biodegradation 17 (2):57-65. doi:10.1007/s10532-005-3050-4

Kolmert Å, Johnson DB (2001) Remediation of acidic waste waters using immobilised, acidophilic sulfate-reducing bacteria. Journal of Chemical Technology & Biotechnology 76 (8):836-843. doi:10.1002/jctb.453

König J, Keesman KJ, Veeken A, Lens PNL (2006) Dynamic Modelling and Process Control of ZnS Precipitation. Separation Science and Technology 41 (6):1025 - 1042

La H-J, Kim K-H, Quan Z-X, Cho Y-G, Lee S-T (2003) Enhancement of sulfate reduction activity using granular sludge in anaerobic treatment of acid mine drainage. Biotechnology Letters 25 (6):503-508. doi:10.1023/a:1022666310393

Labrenz M, Druschel GK, Thomsen-Ebert T, Gilbert B, Welch SA, Kemner KM, Logan GA, Summons RE, Stasio GD, Bond PL, Lai B, Kelly SD, Banfield JF (2000) Formation of Sphalerite (ZnS) Deposits in Natural Biofilms of Sulfate-Reducing Bacteria. Science 290 (5497):1744-1747. doi:10.1126/science.290.5497.1744

Lenz M, Hullebusch EDv, Farges F, Nikitenko S, Corvini PFX, Lens PNL (2011) Combined Speciation Analysis by X-ray Absorption Near-Edge Structure Spectroscopy, Ion Chromatography, and Solid-Phase Microextraction Gas Chromatography−Mass Spectrometry To Evaluate Biotreatment of Concentrated Selenium Wastewaters. Environmental Science and Technology 45 (3):1067-1073

Lettinga G, van Velsen AFM, Hobma SW, de Zeeuw W, Klapwijk A (1980) Use of the upflow sludge blanket (USB) reactor concept for biological wastewater treatment, especially for anaerobic treatment. Biotechnology and Bioengineering 22 (4):699-734. doi:10.1002/bit.260220402

Lewis A, Swartbooi A (2006) Factors Affecting Metal Removal in Mixed Sulfide Precipitation. Chemical Engineering & Technology 29 (2):277-280

Lewis A, van Hille R (2006) An exploration into the sulphide precipitation method and its effect on metal sulphide removal. Hydrometallurgy 81 (3-4):197-204

Lewis AE (2010) Review of metal sulphide precipitation. Hydrometallurgy 104 (2):222-234

Lopes SIC, Sulistyawati I, Capela MI, Lens PNL (2007) Low pH (6, 5 and 4) sulfate reduction during the acidification of sucrose under thermophilic (55 °C) conditions. Process Biochemistry 42 (4):580-591. doi:10.1016/j.procbio.2006.11.004

Mailleret L, Bernard O, Steyer J-P (2004) Nonlinear adaptive control for bioreactors with unknown kinetics. Automatica 40 (8):1379-1385. doi:10.1016/j.automatica.2004.01.030

Maree JP, Strydom WF (1985) Biological sulphate removal in an up-flow packed bed reactor. Water Research 19 (9):1101-1106

Marin P, Alkalay D, Guerrero L, Chamy R, Schiappacasse MC (1999) Design and startup of an anaerobic fluidized bed reactor. . Water Science & Technology 40 (8):63-70

Mersmann A (1999) Crystallization and precipitation. Chemical Engineering and Processing 38 (4-6):345-353

Mishra PK, Das RP (1992) Kinetics of zinc and cobalt sulphide precipitation and its application in hydrometallurgical separation. Hydrometallurgy 28 (3):373-379

Mizuno O, Li YY, Noike T (1998) The behavior of sulfate-reducing bacteria in acidogenic phase of anaerobic digestion. Water Research 32 (5):1626-1634. doi:10.1016/s0043-1354(97)00372-2

Mokone TP, van Hille RP, Lewis AE (2010) Effect of solution chemistry on particle characteristics during metal sulfide precipitation. Journal of Colloid and Interface Science 351 (1):10-18. doi:10.1016/j.jcis.2010.06.027

Nagpal S, Chuichulcherm S, Peeva L, Livingston A (2000) Microbial sulfate reduction in a liquid–solid fluidized bed reactor. Biotechnology and Bioengineering 70 (4):370-380. doi:10.1002/1097-0290(20001120)70:4<370::aid-bit2>3.0.co;2-7

Neculita C-M, Zagury GJ, Bussiere B (2007) Passive Treatment of Acid Mine Drainage in Bioreactors using Sulfate-Reducing Bacteria: Critical Review and Research Needs. J Environ Qual 36 (1):1-16. doi:10.2134/jeq2006.0066

Neculita C-M, Zagury GJ, Bussière B (2008) Effectiveness of sulfate-reducing passive bioreactors for treating highly contaminated acid mine drainage: II. Metal removal mechanisms and potential mobility. Applied Geochemistry 23 (12):3545-3560

Nikolov L, Karamanev D (1987) The inverse fluidization: a new approach to biofilm reactor design to aerobic wastewater treatment. . Paper presented at the Studies in Environmental Science – Environmental Biotechnology,

Okabe S, Nielsen PH, Jones WL, Characklis WG (1995) Rate and s toichiometry of microbial sulfate reduction by *Desulfovibrio-desulfuricans* in biofilms Biofouling 9 (1):63-83

Oleszkiewicz JA, Marstaller T, McCartney DM (1989) Effects of pH on sulfide toxicity to anaerobic processes. Environmental Technology Letters 10 (9):815-822. doi:10.1080/09593338909384801

Oyekola OO, Harrison STL, van Hille RP (2012) Effect of culture conditions on the competitive interaction between lactate oxidizers and fermenters in a biological sulfate reduction system. Bioresource Technology 104 (0):616-621. doi:10.1016/j.biortech.2011.11.052

Pagnanelli F, Cruz Viggi C, Toro L (2010) Isolation and quantification of cadmium removal mechanisms in batch reactors inoculated by sulphate reducing bacteria: Biosorption versus bioprecipitation. Bioresource Technology 101 (9):2981-2987

Papirio S, Villa-Gomez DK, Esposito G, Lens PNL, Pirozzi F (2012) Acid mine drainage treatment in fluidized-bed bioreactors by sulfate-reducing bacteria: a critical review. Critical reviews in environmental science and technology In Press

Parsons JG, Aldrich MV, Gardea-Torresdey JL (2002) Environmental and biological applications of extended X-ray absorption fine structure (EXAFS) and X-ray near structure (XANES) spectroscopies. Applied Spectroscopy Reviews 37 (2):187 - 222

Peters RW, Chang T-K, Ku Y (1984) Heavy metal crystallization kinetics in an MSMPR crystallizer employing sulfide precipitation. Journal Name: AIChE Symp Ser; (United States); Journal Volume: 80:240:Medium: X; Size: Pages: 55-75

Pind P, Angelidaki I, Ahring B, Stamatelatou K, Lyberatos G (2003) Monitoring and Control of Anaerobic Reactors

Biomethanation II. In: Ahring B, Ahring B, Angelidaki I et al. (eds), vol 82. Advances in Biochemical Engineering/Biotechnology. Springer Berlin / Heidelberg, pp 135-182. doi:10.1007/3-540-45838-7_4

Prange A, Modrow H (2002) X-ray absorption spectroscopy and its application in biological, agricultural and environmental research. Reviews in Environmental Science and Biotechnology 1 (4):259-276. doi:10.1023/a:1023281303220

Reis MAM, Almeida JS, Lemos PC, Carrondo MJT (1992) Effect of hydrogen sulfide on growth of sulfate reducing bacteria. Biotechnology and Bioengineering 40 (5):593-600. doi:10.1002/bit.260400506

Ruby MV, Davis A, Nicholson A (1994) In Situ Formation of Lead Phosphates in Soils as a Method to Immobilize Lead. Environmental Science & Technology 28 (4):646-654. doi:10.1021/es00053a018

Ryhiner GB, Heinzle E, Dunn IJ (1993) Modeling and simulation of anaerobic wastewater treatment and its application to control design: Case whey. Biotechnology Progress 9 (3):332-343. doi:10.1021/bp00021a013

Sahinkaya E, Gungor M, Bayrakdar A, Yucesoy Z, Uyanik S (2009) Separate recovery of copper and zinc from acid mine drainage using biogenic sulfide. Journal of Hazardous Materials 171 (1-3):901-906

Samaranayake R, Singhal N, Lewis G, Hyland M (2002) Kinetics of biochemically driven metal precipitation in synthetic landfill leachate. Remediation Journal 13 (1):137-150

Sampaio RMM, Timmers RA, Xu Y, Keesman KJ, Lens PNL (2009) Selective precipitation of Cu from Zn in a pS controlled continuously stirred tank reactor. Journal of Hazardous Materials 165 (1-3):256-265

Sani RK, Peyton BM, Jandhyala M (2003) Toxicity of lead in aqueous medium to Desulfovibrio desulfuricans G20. Environmental Toxicology and Chemistry 22 (2):252-260. doi:10.1002/etc.5620220203

Schubert H, Mersmann A (1996) Determination of Heterogeneous Nucleation Rates. Trans IchemE 78 (Part A)

Schwartz W (1985) Postgate, J. R., The Sulfate-Reducing Bacteria (2nd Edition) X + 208 S., 20 Abb., 4 Tab. University Press, Cambridge 1983. US $ 39.50. Journal of Basic Microbiology 25 (3):202-202. doi:10.1002/jobm.3620250311

Seckler MM, van Leeuwen MLJ, Bruinsma OSL, van Rosmalen GM (1996) Phosphate removal in a fluidized bed—II. Process optimization. Water Research 30 (7):1589-1596. doi:http://dx.doi.org/10.1016/0043-1354(96)00017-6

Shieh W, Keenan J (1986) Fluidized bed biofilm reactor for wastewater treatment

Bioproducts. In, vol 33. Advances in Biochemical Engineering/Biotechnology. Springer Berlin / Heidelberg, pp 131-169. doi:10.1007/BFb0002455

Sierra-Alvarez R, Karri S, Freeman S, Field JA (2006) Biological treatment of heavy metals in acid mine drainage using sulfate reducing bioreactors. Water Science and Technology 54 (2):179-185

Somlev V, Tishkov S (1992) Application of fluidized carrier to bacterial sulphate-reduction in industrial wastewaters purification. Biotechnology Techniques 6 (1):91-96. doi:10.1007/bf02438697

Steed VS, Suidan MT, Gupta M, Miyahara T, Acheson CM, Sayles GD (2000) Development of a Sulfate-Reducing Biological Process To Remove Heavy Metals from Acid Mine Drainage. Water Environment Research 72:530-535

Sukola K, Wang F, Tessier A (2005) Metal-sulfide species in oxic waters Analytica Chimica Acta 528 (2):183–195

Tabak HH, Scharp R, Burckle J, Kawahara FK, Govind R (2003) Advances in biotreatment of acid mine drainage and biorecovery of metals: 1. Metal precipitation for recovery and recycle. Biodegradation 14 (6):423-436

Teitzel GM, Parsek MR (2003) Heavy Metal Resistance of Biofilm and Planktonic Pseudomonas aeruginosa. Applied Environmental Microbiology 69 (4):2313–2320

Tokuda H, Kuchar D, Mihara N, Kubota M, Matsuda H, Fukuta T (2008) Study on reaction kinetics and selective precipitation of Cu, Zn, Ni and Sn with H2S in single-metal and multi-metal systems. Chemosphere 73 (9):1448-1452

Tsukamoto TK, Killion HA, Miller GC (2004) Column experiments for microbiological treatment of acid mine drainage: low-temperature, low-pH and matrix investigations. Water Research 38 (6):1405-1418. doi:10.1016/j.watres.2003.12.012

Vallero MVG, Lettinga G, Lens PNL (2005) High rate sulfate reduction in a submerged anaerobic membrane bioreactor (SAMBaR) at high salinity. Journal of Membrane Science 253 (1–2):217-232. doi:http://dx.doi.org/10.1016/j.memsci.2004.12.032

van Hille RP, A. Peterson K, Lewis AE (2005) Copper sulphide precipitation in a fluidised bed reactor. Chemical Engineering Science 60 (10):2571-2578

van Houten RT, Pol LWH, Lettinga G (1994) Biological sulphate reduction using gas-lift reactors fed with hydrogen and carbon dioxide as energy and carbon source. Biotechnology and Bioengineering 44 (5):586-594. doi:10.1002/bit.260440505

van Hullebusch E, Zandvoort M, Lens P (2003) Metal immobilisation by biofilms: Mechanisms and analytical tools. Reviews in Environmental Science and Biotechnology 2 (1):9-33

Veeken AHM, Akoto L, Hulshoff Pol LW, Weijma J (2003a) Control of the sulfide (S2-) concentration for optimal zinc removal by sulfide precipitation in a continuously stirred tank reactor. Water Research 37 (15):3709-3717

Veeken AHM, de Vries S, van der Mark A, Rulkens WH (2003b) Selective Precipitation of Heavy Metals as Controlled by a Sulfide-Selective Electrode. Separation Science and Technology 38 (1):1 - 19

Villa-Gomez D, Ababneh H, Papirio S, Rousseau DPL, Lens PNL (2011) Effect of sulfide concentration on the location of the metal precipitates in inversed fluidized bed reactors. Journal of Hazardous Materials 192 (1):200-207. doi:10.1016/j.jhazmat.2011.05.002

Villa-Gomez DK, Papirio S, van Hullebusch ED, Farges F, Nikitenko S, Kramer H, Lens PNL (2012) Influence of sulfide concentration and macronutrients on the characteristics of metal precipitates relevant to metal recovery in bioreactors. Bioresource Technology (0). doi:10.1016/j.biortech.2012.01.041

Weijma J, Copini CFM, Buisman CJN, Schultz CE (2002) Biological recovery of metals, sulfur and water in the mining and metallurgical industry. In: In: Water Recycling and Recovery in Industry / Lens, P.N.L., Hulshoff Pol, L.W., Wilderer, P., Asano, T. - London, UK : IWA Publishing, 2002. - (Integrated Environmental Technology Series). - ISBN 1 84339 005 1. pp 605-622

White C, Gadd GM (2000) Copper accumulation by sulfate-reducing bacterial biofilms. FEMS Microbiology Letters 183 (2):313-318

Widdel F (ed) (1988) Microbiology and ecology of sulfate- and sulfur-reducing bacteria. In The Biology of Anaerobic Microorganisms, .

Willow MA, Cohen RRH (2003) pH, Dissolved Oxygen, and Adsorption Effects on Metal Removal in Anaerobic Bioreactors. J Environ Qual 32 (4):1212-1221. doi:10.2134/jeq2003.1212

Zagury GJ, Kulnieks VI, Neculita CM (2006) Characterization and reactivity assessment of organic substrates for sulphate-reducing bacteria in acid mine drainage treatment. Chemosphere 64 (6):944-954. doi:10.1016/j.chemosphere.2006.01.001

CHAPTER
2

EFFECT OF
SULFIDE CONCENTRATION
ON THE LOCATION
OF THE METAL PRECIPITATES
IN THE IFB REACTOR

Abstract

The effect of the sulfide concentration on the location of the metal precipitates within sulfate-reducing inversed fluidized bed (IFB) reactors was evaluated. Two mesophilic IFB reactors were operated for over 100 days at the same operational conditions, but with different chemical oxygen demand (COD) to SO_4^{2-} ratio (5 and 1, respectively). After a start up phase, 10 mg/L of Cu, Pb, Cd and Zn each were added to the influent. The sulfide concentration in one IFB reactor reached 648 mg/L, while it reached only 59 mg/L in the other one. In the high sulfide IFB reactor, the precipitated metals were mainly located in the bulk liquid (as fines), whereas in the low sulfide IFB reactor the metal precipitates were mainly present in the biofilm. The latter can be explained by local supersaturation due to sulfide production in the biofilm. This paper demonstrates that the sulfide concentration needs to be controlled in sulfate reducing IFB reactors to steer the location of the metal precipitates for recovery.

This Chapter has been published as:

Villa-Gomez D, Ababneh H, Papirio S, Rousseau DPL, Lens PNL (2011) Effect of sulfide concentration on the location of the metal precipitates in inversed fluidized bed reactors. Journal of Hazardous Materials 192 (1):200-207.

2.1 Introduction

Shortages in metal and mineral resources are expected in the next decades due to a growing demand by human consumption (Wouters and Bol 2009). Therefore, metal-containing wastewaters can become a resource for metal recovery and reuse. Sulfide precipitation is an efficient method to remove and recover metals from wastewaters (Peters et al. 1984). It is an attractive option over hydroxide and carbonate precipitation due to the lower solubility and faster reaction rates (Bhattacharyya et al. 1981). Biological sulfate reduction has become an attractive alternative method for the production of sulfide. During this process, sulfate reducing bacteria reduce the sulfate to sulfide in the presence of an organic electron donor or hydrogen (Widdel 1988).

Different reactor configurations have been tested for sulfate reduction and metal precipitation (Kaksonen and Puhakka 2007). However, metal recovery cannot always be achieved in these reactors, since metals precipitate partly in the biomass, which hamper their recovery. A promising reactor configuration for metal sulfate reduction and metal removal in a single unit is the inversed fluidized bed (IFB) reactor (Celis et al. 2009). This configuration is based on a floatable carrier material (on which the sulfate reducing bacteria biofilm is formed) which is fluidized downwards, whereas the metal sulfide precipitates settle and thus can be recovered at the bottom of the IFB (Gallegos-Garcia et al. 2009).

From the standpoint of chemistry, supersaturation, which depends on the stoichoimetry of the reactants, is a key factor in understanding the metal sulfide precipitation. Sulfate reduction has been studied to treat organic and inorganic sulfate-rich wastewaters (Lens et al. 2002), and the sulfide concentration varies greatly in these studies. Thus, these lead to different levels of supersaturation. Van Hille et al. (2005) found that high supersaturation causes the rapid precipitation of copper sulfides, often resulting in the formation of fines and hydrated colloidal particles. On the other hand, Lewis and van Hille (2006) found that low sulfide concentrations lead to the formation of aqueous sulfide clusters at high supersaturation points. These conclusions cannot be transferred directly to biological systems, where sulfide is not directly supplied to the reactor, but is produced by the biomass and hence, sulfide supply is linked to the biomass distribution over the reactor.

Bijmans et al. (2009a) investigated the effect of the sulfide concentration on the ZnS precipitation characteristics in a sulfate reducing gas lift reactor. At low sulfide concentrations (0.26 mg/L), larger ZnS particles were formed with better settling properties than at high sulfide concentrations (3.2-70.4 mg/L). These authors used only a small range of sulfide concentration (0.26-70.4 mg/L) and studied the evolution of the particle size of the metal precipitates in a consecutive reactor run. The aim of the present study was to evaluate the effect of a large difference in sulfide concentration (59 versus 501 mg/L) on the fate and recovery efficiency of heavy metals (Cu, Zn, Cd and Zn) in IFB reactors.

2.2 Materials and methods

2.2.1 Source of biomass

The two IFB reactors were inoculated with 25 mL anaerobic sludge from a digester treating activated sludge from a domestic wastewater treatment plant (De Nieuwe Waterweg in Hoek van Holland, The Netherlands). The sludge contained 35.4 g volatile suspended solids (VSS) of the mixed liquor sample per liter of sludge (wet weight). The sludge was added to the reactor, which was kept on recirculation for one day to promote microbial attachment on the carrier material.

2.2.2 Carrier material

The carrier material consisted of 600 mL low-density polyethylene beads (Purell Pe 1810 E, Basell Polyolifins, The Netherlands) of 3 mm diameter. Prior to use, the surface of the polyethylene beads was roughened by abrasion with sand for approximately 15 minutes. Then, the polyethylene beads were rinsed to remove the sand.

2.2.3 Synthetic wastewater

The synthetic wastewater used for the reactor operation and batch experiments contained (mg/L): KH_2PO_4 500, NH_4Cl 200, $CaCl_2 \cdot 2H_2O$ 2500, $FeSO_4 \cdot 7H_2O$ 50 and $MgSO_4 \cdot 7H_2O$ 2500. Lactate was used as electron donor. The pH of the medium was adjusted to 7.0 with NaOH. All reagents were of analytical grade.

2.2.4 IFB reactor

The experiments were conducted in two IFB reactors (Figure 2.1) constructed from a transparent polyvinyl chloride (PVC) pipe operated at room temperature (25° C). Each reactor consisted of a column with a conical bottom of a total volume of 5 L (0.08 m diameter, 1 m height). The flow distributor and gas outlet were mounted in the removable cap covering the reactor. The influent was supplied by using a multichannel peristaltic pump (Watson-Marlow BV, The Netherlands) connected to the influent tank of each reactor. The expansion of the bed (30% of the reactor volume) was maintained by means of the recirculation flow using a magnetic drive pump (IWAKI MD-20R-22ON, Iwaki Holland BV, The Netherlands). The reactor was connected to an equalizer to maintain a constant liquid level in the reactor (Figure 2.1). In addition, the equalizer functioned as a second settler from which metal precipitates were recovered as well.

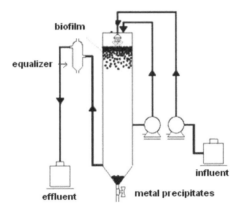

Figure 2.1 Schematic representation of the IFB reactor set-up used in this study.

2.2.5 Experimental design

2.2.5.1 IFB reactor operation

Reactor 1 (R1) and reactor 2 (R2) were run for 109 and 103 days, respectively, at the same operational conditions but with different lactate concentrations (Table 2.1). R1 was operated at a chemical oxygen demand (COD) concentration and COD/SO$_4^{2-}$ ratio (g/g) of 5 g COD/L and 5, respectively, whereas R2 was operated at 1 g COD/L and 1, respectively. After a start up period (I), Cu, Pb, Cd and Zn as chloride salts were added to the influent at a concentration of 5 mg/L each for 7 days (period II). In period III, no metals were added for 5 days to both reactors to ensure the complete removal of metals from the liquid phase prior to the increase of the metal concentration to 10 mg/L each (period IV to V).

Table 2.1 Operational conditions of the two IFB reactors.

Parameter		\multicolumn{5}{c}{Experimental periods}				
		I	II	III	IV	V
Characteristic		Start up	Metal adaptation	No metals	Metal removal test	Decrease of HRT
Days	R1	0-43	44-50	51-56	57-76	77-109
	R2	0-35	36-42	43-48	49-66	67-103
HRT (days)		1	1	1	1	0.37
Metals added* (mg/L)		No	5	No	10	10

T=Room temperature (25 C), pH= 7, * Cu, Zn, Pb and Cd.

From period I to IV, the HRT was maintained at 24 h, which was decreased to 9 h after day 76 for R1 and day 66 for R2 (period V) to test the robustness of the systems with respect to the metal removal efficiencies. The amount of COD, sulfate and dissolved metals removed were measured in the effluent during the whole reactor operation, whereas acetate and dissolved sulfide were started to be measured when the metal removal test started (period IV).

It should be note that the dissolved sulfide concentration was always maintained above the stoichoimetric levels ($[S]/[\sum M^{2+}]$ mol/mol) during the metal removal periods in order to ensure metal sulfide precipitation (confirmed by Visual Minteq Version 3.0, US EPA, 1999, http://www.lwr.kth.se/English/OurSoftware/vminteq/index.html) besides the presence of other possible precipitants contained in the synthetic wastewater. Iron was also considered in the $[S]/[\sum M^{2+}]$ calculation since the medium contained rather high concentrations of this metal that easily precipitates as sulfide.

2.2.5.2 Batch experiments

Batch experiments to quantify the sulfide production in the presence and absence of metals were performed at room temperature (25 °C) in serum bottles of 117 mL shaken at 100 rpm. The bottles contained 5 mL of carrier material withdrawn from R1 at the end of the reactor operation and 112 mL synthetic wastewater.

2.2.5.2.1 Sulfate-reducing activity

The sulfate reducing activity (SRA) was determined in six serum bottles using lactate (1 g COD/L) as the substrate at a COD/SO$_4^{2-}$ ratio of 1 (Exp A). Prior to the experiment, the carrier material with the biofilm was stored at 4 °C. Therefore, the biofilm was activated to the experimental conditions of Exp A, but at 30 °C for 72 hours. After this, the serum bottles were refilled with fresh synthetic wastewater containing lactate.

2.2.5.2.2 Metal precipitation with active biofilm

In order to study the effect of the metals on the sulfate reducing activity, three of the serum bottles used in Exp A, were refilled with fresh synthetic wastewater upon finishing (Exp B) and the other three were kept with the same medium (adding extra synthetic wastewater to cover the headspace) (Exp C). Cu, Zn, Cd and Pb were added to the serum bottles to an initial concentration of 10 mg/L.

The sulfide concentration was determined in Exp A, B and C approximately every 4 hours and the metal concentration every hour during the initial 24 hours of Exp B and C.

2.2.6 Analysis

Total suspended solids (TSS) and volatile suspended solids (VSS) in the biofilm are reported per gram of dry polyethylene and were determined according to standard methods (APHA 2005) after detaching the biofilm from the polyethylene by successive washings with deionized water in an ultrasonic bath. COD was determined by the close reflux method (APHA 2005). Acetate was measured by gas chromatography (GC-CP 9001 Chrompack) after acidification of the samples with 5% concentrated formic acid and filtration through a 0.45 µm nitrocellulose filter (Millipore). The gas chromatograph was fitted with a WCOT fused silica column, the injection and detector temperatures were 175 and 300 °C, respectively. The temperature of the oven was kept at 115 °C. The carrier gas was helium at 100 mL/min.

Sulfide was determined spectrophotometrically by the colorimetric method described by Cord-Ruwisch (1985) using a spectrophotometer (Perkin Elmer Lambda20). Sulfate was measured with an ion chromatograph (ICS-1000 Dionex with ASI-100 Dionex). The column (IonPac AS14n) was used in the ion chromatograph at a flow rate of 0.5

mL/min with an 8 mM Na_2CO_3/1 mM $NaHCO_3$ eluent, a temperature of 35 °C, a current of 35 mA, an injection volume of 10 μL and a retention time of 8 min.

Metals were measured by flame (AAS Perkin Elmer 3110) and furnace (AAS Solaar MQZe GF95) spectroscopy. Metal samples from batch experiments to determine the metal precipitation rate were measured in the liquid phase after diluting, acidifying with 5% HNO_3 and passing the sample through a 0.45 μm nitrocellulose filter (Millipore). Metal samples from the bottom of the reactor and equalizer were analyzed after taking 50 mL of the liquid containing the precipitates and acidifying with 20% HNO_3 to ensure the complete dissolution of the metal precipitates. After this, the procedure for metal measurements mentioned above was followed.

2.2.7 Calculations

The metal precipitates that could not settle at the bottom of the reactor and equalizer due to their small size were defined as fines. It is important to consider these fines to prove the system not only for the removal of metals but also for the potential recovery. The dissolved metal concentration (excluding fines) was determined by passing the liquid samples through a filter (0.45 μm) prior to acidification.

The metal removal efficiency in the down-flow FBR was defined as:

$$\text{Metal removal efficiency (\%)} = \frac{M_{in} - M_{out,dissolved}}{M_{in}} \times 100$$

M_{in} = Metal concentration in the feed (mg/L)

$M_{out,total}$ = Dissolved metal concentration in the outlet (mg/L)

The results of the metals accumulated in the bottom of the reactor and equalizer during each experimental period were used to calculate the metal recovery efficiency. Then, mass balance calculations were done to determine the fate of the metals. The metal recovery in the IFB reactor was defined as:

$$\text{Metal recovery efficiency (\%)} = \frac{M_{total} - \sum_{1}^{n}\left(M_{eq} + M_{b}\right)}{M_{total}} \times 100$$

M_{eq} = Metals (mg) from the equalizer in the sample

M_{b} = Metals (mg) from the bottom of the reactor in the sample

M_{total} =Total metals (mg) in the influent that entered the reactor over an operational

period.

n = Number of samples in each operational period

2.3 Results

2.3.1 IFB reactor operation

Table 2.2 compares the effluent characteristics of R1 and R2. The performance of both

reactors during the start up period was characterized by instabilities in COD and sulfate

removal. The gradual increase of the sulfate removal efficiency and the change in color

of the polyethylene (Figure 2.2) and the reactor liquids from brown to black confirmed

that anaerobic conditions and a sulfate reducing biofilm developed. The pH of the

effluent was lower than the influent pH (6.6) in R1, while the pH was maintained at 7.0

in R2.

In period II, sulfate removal efficiencies were on average 88% and 68%, and the

average pH was 7.1 and 7.6 in R1 and R2, respectively, with no changes in the

subsequent periods. The COD removal efficiency in R1 and R2 increased to 22% and

48%, respectively.

Figure 2.2 Picture of the polyethylene beads a) before and b) after the biofilm formation

in the IFB reactors.

Table 2.2 Effluent characteristics and removal efficiencies for sulfate reduction during the operation of the IFB reactors (mean ± standard deviation).

	Experimental periods				
	I^a	$II^{a, c}$	III^a	$IV^{a, d}$	$V^{b, d}$
R1					
COD removal efficiency (%)	15 (±3)	22*	34*	27 (±7)	27 (±10)
SO_4^{2-} removal efficiency (%)	56 (±24)	88(±3)	70*	76(±15)	74 (±13)
Acetate concentration (g COD/L)	ND	ND	ND	1.39 (±0.6)	1.36 (±0.4)
Sulfide concentration (mg/L)	ND	ND	ND	212 (±26)	648 (±153)
$[S]/[\sum M^{2+}]$ (mol/mol)	ND	ND	ND	14.8	45.15
Effluent pH	6.6 (±0.2)	7.1 (0)	7.0 (0)	7.1 (±0.1)	7 (±0.2)
R2					
COD removal efficiency (%)	35 (±8)	48*	-	68 (±11)	53 (±18)
SO_4^{2-} removal efficiency (%)	59 (±14)	68 (±15)	53*	17 (±11)	38 (±17)
Acetate concentration (g COD/L)	ND	ND	ND	0	0
Sulfide concentration (mg/L)	ND	ND	ND	59 (±24)	44 (±30)
$[S]/[\sum M^{2+}]$ (mol/mol)	ND	ND	ND	4.11	5.16
Effluent pH	7.0 (±0.1)	7.6 (0)	7.6 (±0.1)	7.5 (±0.2)	7.5 (±0.1)

HRT= [a]24 h and [b]9 h, Influent metal conc. = [c]5 mg/L and [d]10 mg/L, *Average of two values, NR= Not determined, $[\sum M^{2+}]$: The sum of Zn, Cu, Pb, Cd and Fe molar concentrations.

In period III, the sulfate removal efficiency decreased in both reactors, in R1 to 70% and in R2 to 53%, while the COD removal efficiency continued to increase to 34 % in R1. In period IV, a recovery in the sulfate removal efficiency was observed in R1 (76%), while in R2 the sulfate removal efficiency dropped to 17%. The COD removal efficiency continued increasing to 68% in R2, while it decreased to 27% in R1. In this period, it was shown that part of the COD supplied was transformed to acetate in R1 (1.39 g COD/L), while the acetate concentration in R2 was below the detection limit. During period IV, the mean sulfide concentration in R1 and R2 was 212 mg/L and 59 mg/L, respectively. In the same period, polyethylene beads started to settle in both reactors, this was more pronounced in R2. This caused a failure of the recirculation pump on day 87 in R1 and on day 66, 81 and 101 in R2. The failure of the pump was also reflected in the lower sulfate removal efficiency by R2 on days 74, 88, and 102 (Figure 2.3).

In period V, the change of the HRT from 24 to 9 h did not vary the COD nor sulfate
removal efficiencies in R1, while in R2 a partial recovery of the sulfate removal
efficiency was observed (38%). The maximum sulfide concentration was reached in
period V for R1 (648 mg/L), while only 44 mg/L of sulfide was produced on average in
R2. In period V, it was confirmed by the mass balances (data not shown) that most of
the COD consumed was used for sulfate reduction (>87%) in R1. Acetate production
remained close to the value obtained in the previous period (1.36 g/L) and accounted for
43.4% of the COD consumed.

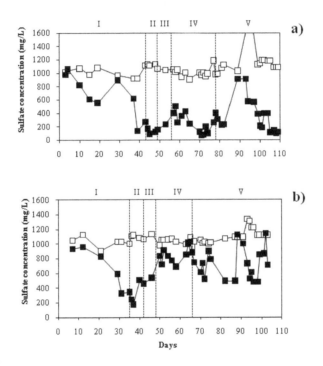

Figure 2.3 Evolution of the sulfate concentration in the influent (□) and effluent (■)
during reactor operation of a) R1 and b) R2.

2.3.1.2 Metal removal/recovery in the IFB reactors

Table 2.3 shows the metal removal efficiency of both R1 and R2 for the periods where
the metals were added to the influent (II, IV and V). Metal removal efficiencies
exceeded 91% in both reactors in period II, which further increased for both reactors
and exceeded 95% in period IV. The Cu, Zn, Pb and Cd removal efficiencies in both

reactors exceeded 98.4%, 96.5%, 96%, and 97.9%, respectively in period V, but were slightly higher in R2, in spite of the higher sulfide concentration in R1 compared to R2. Neither the increment of the metal concentration (Period IV) nor the change in HRT affected the metal removal efficiency (Period V).

Table 2.3 also shows the percentage of metals that had accumulated at the bottom of the reactor and equalizer, expressed as the metal recovery efficiency. In period II, only less than 5% of the metals were recovered in R2, while in R1 up to 17% of the metals could be recovered. In Period IV, the metal recovery increased in both reactors, the highest values were obtained in R2 (29.8%, 26.9%, 30.0%, and 26.2% for Cu, Zn Pb and Cd respectively). The recovery continued to increase in both reactors in period V achieving values of 41.1-60.3%. In general, the difference in the sulfide concentration and the operational conditions of both reactors did not result in any relevant variation in metal recovery.

Table 2.3 Metal removal efficiency and metal recovery in the IFB reactors in periods II, IV and V (mean ± standard deviation).

Period	Metal removal efficiency (%)			Metal recovery* (%)		
	II	**IV**	**V**	**II**	**IV**	**V**
R1						
Cu	99.1 (±1.3)	97.1 (±2.9)	98.4 (±3.1)	16.8	17.2	49.4
Zn	92.3 (±4.2)	95.1 (±4.4)	96.5 (±3.8)	13.6	16.5	43.6
Pb	92.7 (±4.9)	97.3 (±1.9)	96.0 (±5.1)	13.2	21.4	57.9
Cd	95.6 (±1.7)	95.2 (±2.5)	97.9 (±3.0)	15.7	17.2	46.3
R2						
Cu	99.1 (±1.5)	96.7 (±3.3)	99.9 (±0.3)	4.1	29.8	41.1
Zn	91.3 (±10.0)	95.6 (±2.3)	98.6 (±1.2)	2.7	26.9	44.2
Pb	92.4 (±8.0)	96.1 (±3.4)	99.2 (±1.1)	3.5	30.0	60.3
Cd	96.5 (±2.4)	95.3 (±2.9)	99.7 (±0.3)	2.9	26.0	47.4

* Metal recovery from bottom of the reactor and equalizer.

2.3.1.3 Location of metal precipitation in the biofilm

It was assumed that the metals which were not present in the effluent, bottom of the reactor and equalizer were present in the reactor as fines or adsorbed or precipitated in the biofilm. At the end of the reactor operation, metals, TSS and VSS in the biofilm were analyzed on polyethylene bead samples located at the top and bottom of the polyethylene bed (Table 2.4).

Table 2.4 TSS, VSS and metal concentration in the biofilm of both IFB reactors at the end of the experiment (average of duplicate samples). Samples taken at the top and the bottom of the polyethylene bed.

| | | TSS | VSS | Metal concentration ($mg/g_{polyethylene}$) | | | |
		$(mg/g_{polyethylene})$		Cu	Pb	Zn	Cd
R1	Top	2.7	0.8	0.4	0.2	0.5	0.3
	Bottom	6.8	1.4	0.4	0.2	0.5	0.3
R2	Top	4.8	0.8	0.6	0.3	0.8	0.7
	Bottom	35.6	4.6	1.5	0.8	1.8	1.4

The TSS and VSS in the biofilm were different at the top and at the bottom of both fluidized beds. The TSS was especially high at the bottom of the R2 polyethylene bed. No differences in the metal concentration were observed between beads sampled at the top and the bottom of the fluidized bed in R1. In general, higher metal concentrations in the biofilm were found in R2. Moreover, the differences in metal concentration between the top and the bottom were significantly higher in R2.

2.3.2 Batch experiments

2.3.2.1 Sulfate reducing activity of the biofilm

Figure 2.4 compares the sulfide production in the absence (Exp A) and presence of metals (Exp B). The sulfide started to increase after a lag phase of 16 hours for Exp A, while no lag phase was observed for Exp B. Prior to the addition of metals (Exp A), the SRA was 0.16 mg S^{2-}/mg VSS-h, which increased to 0.21 mg S^{2-}/mg VSS-h after the addition of metals (Exp B). These results confirm that the metal concentration itself was not directly inhibitory to the biofilm.

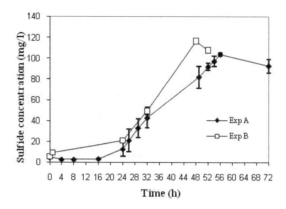

Figure 2.4 Sulfide production in the absence (Exp A) and in the presence (Exp B) of metals in the batch experiments with the carrier material withdrawn from R1 (Vertical bars show ± standard deviation of three replicates).

2.3.2.2 Metal precipitation with active biofilm

Figure 2.5 compares metal precipitation when sulfide is formed by the sulfate reducing biofilm during the batch experiment (Exp B) and when sulfide is already produced by the sulfate reducing biofilm at the start of the experiment (Exp C). In Exp B, the metal concentration decreased within the first hour, even when sulfide was not yet accumulating in the medium. After 24 hours, the sulfide concentration was 30.9 mg/L, at this sulfide concentration Pb could not be detected in the liquid phase anymore (Figure 2.5). The sulfide concentration continued increasing with time, whereas the metal concentration reached steady state after 32 hours. Metals remained in the liquid phase at 1 mg/L for Cu and 0.3 mg/L for Cd and Zn.

In Exp C, almost all the Cu, Cd and Pb precipitated within the first hour, while 6.7 mg/L of the Zn remained in the liquid phase. After 24 hours, Cd and Pb were not found in the liquid phase anymore, while 1.2 mg/L of Zn and 2.1 mg/L of Cu remained in the liquid phase with slight variations during the subsequent 48 hours. After 52 hours, Cu and Zn had further decreased and only 0.9 mg/L Zn remained in the liquid phase.

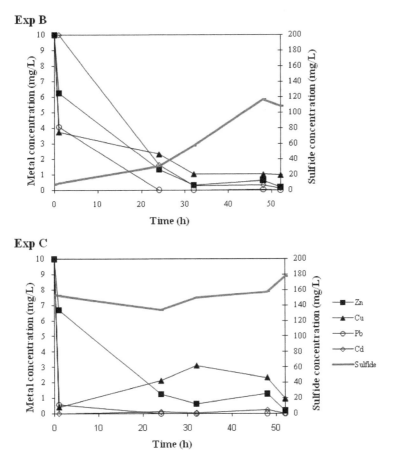

Figure 2.5 Evolution of the metal concentration in time when sulfide is formed by the biofilm (Exp B) or when sulfide (152 mg/L) is present in the liquid phase from t=0 onwards (Exp C) in the liquid phase (top) and in the biofilm (bottom) (average of duplicate samples).

2.4 Discussion

This paper shows that the sulfide concentration in the reactor mixed liquor controls the location of the metal precipitates in an IFB reactor. Moreover, the results suggest that the recovery of the metal precipitates at the bottom of the IFB reactors is independent of the sulfide concentration, and hence, other mechanisms determine the settling properties of the sulfide precipitates, e.g. agglomeration. To the best of our knowledge, the present study is the first to report the location of metal sulfide precipitation as a function of the sulfide concentration in an IFB reactor treating a multimetal wastewater.

2.4.1 Location of metal sulfide precipitation

Table 2.4 shows that the metals are partly immobilized in the biofilm. Bijmans et al. (2009b) suggested that biofilms might function as nucleation seeds, enhancing the crystal growth for metal sulfides. Other authors have induced the precipitation of heavy metals on the sand surface in fluidized bed reactors (FBR's) using sulfide or carbonate (Zhou et al. 1999; Guillard and Lewis 2001; van Hille et al. 2005). Zhou et al. (1999) observed that when the ratio of carbonate to metals (Cu, Ni and Zn) was low, metal precipitation was coated on the sand surface, while at high ratios (6:1 and 3:1) the precipitation occurred in the bulk solution. These findings are in agreement with this study: in R2 the sulfide concentration in the bulk liquid was much lower than in R1. Therefore, supersaturation, and thus precipitation of metal sulfides, mainly occurred within the biofilm in R2, where the sulfide is produced. In R1, in contrast, a larger fraction of the supersaturation, and thus precipitation, occurred in the bulk liquid, due to the much higher sulfide concentration in the reactor mixed liquor, resulting in a lower metal content in the biofilm (Table 2.4 and Figure 2.6).

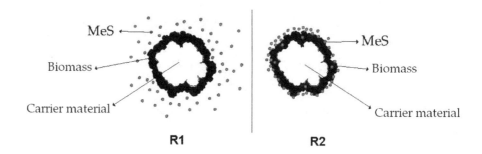

R1 **R2**

Figure 2.6 Location of the metal sulfide (MeS) precipitate formation, depending on the sulfide concentration: MeS precipitation outside the biofilm at high sulfide concentrations as in R1 and MeS precipitation in the biofilm at low sulfide concentrations as in R2.

Metals might also be immobilized in biofilms due to sorption onto the microorganisms and/or on their extracellular polymeric substances (EPS) (Beech and Cheung 1995;

White and Gadd 2000; Chen et al. 2000). However, this mechanism was apparently not the case in this study as the sulfide concentration was always maintained above the stoichoimetric value to precipitate the four metals as metal sulfides in both reactors. In addition, the metal sulfide formation rate is extremely fast compared to the adsorption mechanisms reported in the literature (Peters et al. 1984; Wang et al. 2001). Wang et al. (2001) showed that CdS was formed at the cell surface because the cadmium sulfide formation rate is extremely fast (K_{CdS}= 3.7 × 10^5 s^{-1}) compared to the slow rate of sulfide transport (K_{diff}= 1000 s^{-1}) on the bacterial cell surface.

2.4.2 Metal recovery

In both reactors, the metal removal and recovery did not vary with the sulfide concentration or the different operational conditions. Metal removal efficiencies were on average 95% for the four metals tested (Table 2.3), while metal recovery from the bottom of the reactor and equalizer was less than 50% in both reactors (Table 2.3). In previous studies, similar metal removal efficiencies have been achieved in biological reactors treating wastewaters with more than two metals in a single unit (Gallegos-Garcia et al. 2009; Kaksonen et al. 2003; Kaksonen et al. 2004; Sierra-Alvarez et al. 2006; Steed et al. 2000; La et al. 2003; Foucher et al. 2001), although, the recovery of metals is often not reported. Gallegos-Garcia et al. (2009) reported 76-97% recovery of Fe, Zn, and Cd in an IFB at initial metal concentrations between 5 and 320 mg/L and sulfide concentrations over 140 mg/L. These results differ considerably from this study. The difference could be related to the treatment of the samples. Gallegos-Garcia et al. (2009) as well as other authors (Kaksonen et al. 2003) have reported the metal recovery assuming that the TSS concentrations equal the metal sulfide composition. However, in this study it was shown that the TSS also contained salts from the mineral medium that contribute to the weight (observed in the results of the biofilm composition, Table 2.4). Therefore, this study reports the recovery of the metals based on the direct measurement of metals after acidification of the recovered solids (Table 2.3).

2.4.2.1 Formation of fines

The difference between the metal removal and recovery efficiency (Table 2.3) can be party explained by the accumulation of metals in the biofilm (Table 2.4), but also by the formation of small precipitates (<0.45 μm), known as fines, which do not settle at the bottom of the reactor and leave the IFB reactor with the effluent. These fines did not contribute to the metal concentrations in the effluent, as samples were filtered through a 0.45 μm filter prior to analyses.

The formation of fines is attributed to high levels of supersaturation (Guillard and Lewis 2001) and high nucleation rates (Veeken et al. 2003) of the metal sulfides. This was observed in the batch experiments (Figure 2.5): when the sulfide was slowly produced (Exp B), the precipitation rate could be quantified, whereas when sulfide was already present (Exp C), metal precipitation occurs too fast to monitor the metal precipitation kinetics of Cu, Pb and Cd. Note that the ZnS precipitation is slower due to the higher solubility product compared with the solubility products of the other three metals tested (Sampaio et al. 2009).

2.4.2.2 Effect of agglomeration

The metal recovery gradually increased in both reactors during reactor operation (Table 2.3), despite the differences in operational conditions including sulfide and acetate production (Table 2.2). Both parameters influence the size of the particles, which affects the settling properties of the metal precipitates for recovery. However, this contribution is of lesser importance since the size range of the particles is still very small to allow fast settling (Lewis 2010).

Fines in the reactor can form agglomerates, that later settle and dewater better. Precipitation occurs through several steps: nucleation, crystal growth, and eventually agglomeration (Al-Tarazi et al. 2004). Large particles can be produced if the supersaturation is optimum to promote crystal growth, and the residence time of the crystals is long enough to promote agglomeration (Mersmann 1999). In the precipitation process of metal sulfides, which have an extremely low solubility (Sampaio et al. 2009), the formation of fines is predominant over crystal formation. Therefore, agglomeration is an important mechanism for the settling of the precipitates. Biological processes can contribute to the agglomeration mechanism. It has been shown that the presence of extracellular proteins promotes the aggregation of metal sulfide nanoparticles (Moreau et al. 2007). Further research using Scanning Electron Microscope and Transmission Electron Microscope (TEM) coupled to X-ray diffraction (XRD) and dewatering tests are required to relate the morphology of the produced solids in the IFB reactor with substances present in sulfate reducing bioreactors, for example, extracellular proteins or nutrients added for bacterial growth.

2.4.2.3 Reactor operation

A period of approximately 40 days was necessary to form a sulfate reducing biofilm on the polyethylene beads in an IFB reactor using lactate as electron donor, independent of the COD/SO$_4^{2-}$ ratio. A similar time was necessary in the study of Celis et al. (2009), who used a lactate:ethanol mixture, a COD/SO$_4^{2-}$ ratio of 0.6 and operated in batch mode for 45 days.

In R1, the production of acetate by incomplete oxidizers was probably more favorable than in R2 due to the excess of COD (Oyekola et al. 2009). This acetate also explains why the pH level did not increase in R1 (Table 2.2), since the incomplete oxidation of lactate generates protons (Kaksonen et al. 2003). Contrary to R1, in R2, acetate was not detected during the reactor operation and the pH increased in the effluent by H$_2$S and CO$_2$ production (Table 2.2). This suggests that different microbial populations developed in both reactors.

The comparison of Exp A and Exp B in the batch experiments (Figure 2.5) confirmed that the metal concentration itself was not directly inhibitory to the biofilm. The inhibitory concentrations to SRB reported for Cu, Zn, Pb and Cd are in the range of 6-100, 13–65, 25->80 and >4–112 mg/L, respectively (Kaksonen and Puhakka 2007). The inhibitory effects of metals, however, depend on the experimental conditions (Hao et al. 1994), e.g. time exposure, which differed in batch experiments and continuous reactor operation.

2.5 Conclusion

Several studies with sulfate reducing bioreactors to treat metal containing streams have mainly concentrated on maximizing the sulfate reduction rate. In the present study, it was shown that the sulfide concentration is important, not only to precipitate metals but also to steer the location where the precipitates are formed. Moreover, a sulfide concentration in excess is not desirable since it causes residual pollution problems in the effluents, and may impose a sulfide removal post-treatment step.

References

Al-Tarazi M, Heesink ABM, Azzam MOJ, Yahya SA, Versteeg GF (2004) Crystallization kinetics of ZnS precipitation; an experimental study using the mixed-suspension-mixed-product-removal (MSMPR) method. Crystal Research and Technology 39 (8):675-685. doi:10.1002/crat.200310238

APHA APHA (2005) Standard methods for examination of water and wastewater. 20 edn., Washington D.C.

Beech IB, Cheung CWS (1995) Interactions of exopolymers produced by sulphate-reducing bacteria with metal ions. International Biodeterioration & Biodegradation 35 (1-3):59-72

Bhattacharyya D, Jumawan AB, Sun G, Sund-Hagelberg C, Schwitzgebel K (1981) Precipitation of heavy metals with sodium sulfide: Bench-scale and full-scale experimental results. Paper presented at the AIChE Symposium Series,

Bijmans MFM, van Helvoort P-J, Buisman CJN, Lens PNL (2009a) Effect of the sulfide concentration on zinc bio-precipitation in a single stage sulfidogenic bioreactor at pH 5.5. Separation and Purification Technology 69 (3):243-248

Bijmans MFM, van Helvoort P-J, Dar SA, Dopson M, Lens PNL, Buisman CJN (2009b) Selective recovery of nickel over iron from a nickel-iron solution using microbial sulfate reduction in a gas-lift bioreactor. Water Research 43 (3):853-861

Celis L, Villa-Gómez D, Alpuche-Solís A, Ortega-Morales B, Razo-Flores E (2009) Characterization of sulfate-reducing bacteria dominated surface communities during start-up of a down-flow fluidized bed reactor. Journal of Industrial Microbiology and Biotechnology 36 (1):111-121

Cord-Ruwisch R (1985) A quick method for the determination of dissolved and precipitated sulfides in cultures of sulfate-reducing bacteria. Journal of Microbiological Methods 4 (1):33-36

Chen B-Y, Utgikar VP, Harmon SM, Tabak HH, Bishop DF, Govind R (2000) Studies on biosorption of zinc(II) and copper(II) on Desulfovibrio desulfuricans. International Biodeterioration & Biodegradation 46 (1):11-18

Foucher S, Battaglia-Brunet F, Ignatiadis I, Morin D (2001) Treatment by sulfate-reducing bacteria of Chessy acid-mine drainage and metals recovery. Chemical Engineering Science 56 (4):1639-1645

Gallegos-Garcia M, Celis LB, Rangel-Méndez R, Razo-Flores E (2009) Precipitation and recovery of metal sulfides from metal containing acidic wastewater in a sulfidogenic down-flow fluidized bed reactor. Biotechnology and Bioengineering 102 (1):91-99

Guillard D, Lewis AE (2001) Nickel Carbonate Precipitation in a Fluidized-Bed Reactor. Industrial & Engineering Chemistry Research 40:5564-5569

Hao OJ, Huang L, Chen JM, Buglass RL (1994) Effects of metal additions on sulfate reduction activity in wastewaters. Toxicological & Environmental Chemistry 46 (4):197-212. doi:10.1080/02772249409358113

Kaksonen AH, Franzmann PD, Puhakka JA (2004) Effects of hydraulic retention time and sulfide toxicity on ethanol and acetate oxidation in sulfate-reducing metal-precipitating fluidized-bed reactor. Biotechnology and Bioengineering 86 (3):332-343

Kaksonen AH, Puhakka JA (2007) Sulfate Reduction Based Bioprocesses for the Treatment of Acid Mine Drainage and the Recovery of Metals. Engineering in Life Sciences 7 (6):541-564

Kaksonen AH, Riekkola-Vanhanen ML, Puhakka JA (2003) Optimization of metal sulphide precipitation in fluidized-bed treatment of acidic wastewater. Water Research 37 (2):255-266

La H-J, Kim K-H, Quan Z-X, Cho Y-G, Lee S-T (2003) Enhancement of sulfate reduction activity using granular sludge in anaerobic treatment of acid mine drainage. Biotechnology Letters 25 (6):503-508

Lens P, Vallerol M, Esposito G, Zandvoort M (2002) Perspectives of sulfate reducing bioreactors in environmental biotechnology. Reviews in Environmental Science and Biotechnology 1 (4):311-325. doi:10.1023/a:1023207921156

Lewis A, van Hille R (2006) An exploration into the sulphide precipitation method and its effect on metal sulphide removal. Hydrometallurgy 81 (3-4):197-204

Lewis AE (2010) Review of metal sulphide precipitation. Hydrometallurgy 104 (2):222-234

Mersmann A (1999) Crystallization and precipitation. Chemical Engineering and Processing 38 (4-6):345-353

Moreau JW, Weber PK, Martin MC, Gilbert B, Hutcheon ID, Banfield JF (2007) Extracellular Proteins Limit the Dispersal of Biogenic Nanoparticles. Science 316 (5831):1600-1603. doi:10.1126/science.1141064

Oyekola OO, van Hille RP, Harrison STL (2009) Study of anaerobic lactate metabolism under biosulphidogenic conditions. Water Research 43: 3345-3354

Peters RW, Chang T-K, Ku Y (1984) Heavy metal crystallization kinetics in an MSMPR crystallizer employing sulfide precipitation. Journal Name: AIChE Symp Ser; (United States); Journal Volume: 80:240:Medium: X; Size: Pages: 55-75

Sampaio RMM, Timmers RA, Xu Y, Keesman KJ, Lens PNL (2009) Selective precipitation of Cu from Zn in a pS controlled continuously stirred tank reactor. Journal of Hazardous Materials 165 (1-3):256-265

Sierra-Alvarez R, Karri S, Freeman S, Field JA (2006) Biological treatment of heavy metals in acid mine drainage using sulfate reducing bioreactors. Water Science and Technology 54 (2):179-185

Steed VS, Suidan MT, Gupta M, Miyahara T, Acheson CM, Sayles GD (2000) Development of a Sulfate-Reducing Biological Process To Remove Heavy Metals from Acid Mine Drainage. Water Environment Research 72:530-535

van Hille RP, A. Peterson K, Lewis AE (2005) Copper sulphide precipitation in a fluidised bed reactor. Chemical Engineering Science 60 (10):2571-2578

Veeken AHM, de Vries S, van der Mark A, Rulkens WH (2003) Selective Precipitation of Heavy Metals as Controlled by a Sulfide-Selective Electrode. Separation Science and Technology 38 (1):1 - 19

Wang CL, Clark DS, Keasling JD (2001) Analysis of an engineered sulfate reduction pathway and cadmium precipitation on the cell surface. Biotechnology and Bioengineering 75 (3):285-291. doi:10.1002/bit.10030

White C, Gadd GM (2000) Copper accumulation by sulfate-reducing bacterial biofilms. FEMS Microbiology Letters 183 (2):313-318

Widdel F (ed) (1988) Microbiology and ecology of sulfate- and sulfur-reducing bacteria. In The Biology of Anaerobic Microorganisms, .

Wouters H, Bol D (2009) Material scarcity. Stichting Materials innovation institute (M2i), Delft, The Netherlands

Zhou P, Huang J-C, Li AWF, Wei S (1999) Heavy metal removal from wastewater in fluidized bed reactor. Water Research 33 (8):1918-1924

CHAPTER

3

INFLUENCE OF

MACRONUTRIENTS

ON THE METAL SULFIDE

PRECIPITATE

CHARACTERISTICS

Abstract

Metal sulfide recovery in bioreactors highly depends on their purity and settling properties. In this study, the influence of macronutrients and biofilm on the Cu, Pb, Cd and Zn depletion kinetics and characteristics was evaluated in batch experiments with sulfide at different concentrations. The kinetic values from the plots of the metals followed the solubility product order of the metal sulfides. They also showed that at low sulfide concentrations, the metals with slower depletion rates are susceptible to other removal mechanisms as well, i.e. biosorption onto the biofilm and precipitation with the macronutrients. X-ray absorption spectroscopy showed that the main mechanism of Zn removal is its sorption onto apatite $(Ca_5(PO_4)_3^+(OH^-)$, a compound formed due to the addition of macronutrients. The size of the precipitates was similar in all experiments regardless the sulfide concentration (8.1-10.0 μm).

This Chapter has been published as:

Villa-Gomez DK, Papirio S, van Hullebusch ED, Farges F, Nikitenko S, Kramer H, Lens PNL (2012) Influence of sulfide concentration and macronutrients on the characteristics of metal precipitates relevant to metal recovery in bioreactors. Bioresource Technology (0). doi:10.1016/j.biortech.2012.01.041

3.1 Introduction

Metal recovery and reuse from industrial wastewaters is beneficial for both economic and environmental reasons. Sulfide precipitation (Equation 1) offer several advantages over other precipitation mechanisms such as high degree of metal removal over a broad pH range and better thickening and dewatering characteristics (Fu and Wang 2011).

$$H_2S + M^{2+} \rightarrow MS_{(s)} + 2 H^+ \qquad (1)$$
where M^{2+} = metal, such as Zn^{2+}, Cu^{2+}, Pb^{2+} and Ni^{2+}

Furthermore, sulfide gives the possibility for selective precipitation of metals due to differences in solubility products of the different metal sulfides (Sampaio et al. 2009). Typical sulfide sources used to precipitate metals are Na_2S, NaHS and H_2S (Peters et al. 1984). Biogenic sulfide produced by sulfate reducing bacteria (SRB) is also an option for metal precipitation from wastewater, especially when besides metals, also sulfate is a major wastewater constituent (Equation 2), e.g. in wastewaters from metal refineries and acid mine drainage. The biological sulfate reducing process requires the supply of an electron donor and carbon source as well as essential nutrients to maintain microbial activity:

$$2 CH_2O + SO_4^{2-} \rightarrow H_2S + 2 HCO_3^- \qquad (2)$$
where CH_2O = electron donor

Reactor configurations studied for sulfate reduction and metal removal in a single unit are UASB reactors, UAPBR reactors, fixed-bed reactor, FBRs, and the inversed fluidized bed (IFB) reactor (for reviews see (Kaksonen and Puhakka 2007). The later is based on a floatable carrier material fluidized downwards on which the sulfate reducing bacteria biofilm is formed, whereas metal sulfide precipitates can be recovered at the bottom of the reactor (Villa-Gomez et al. 2011; Gallegos-Garcia et al. 2009).

Metal sulfide precipitation and recovery not only depends on the reactor configuration but also on the settling properties of the metal precipitates as well as the influence of other metal removal mechanisms involved. With respect to the settling properties, the separation of sulfide precipitates has proven difficult due to the formation of small, poorly settling particles, or even sulfide clusters at high supersaturation conditions

(Lewis and Swartbooi 2006), for example inside the sulfate reducing biofilm (Villa-Gomez et al. 2011). Furthermore, substances commonly present in the bioreactor mixed liquor affect the size and structure of the metal precipitates (Esposito et al. 2006; Sampaio et al. 2009; Bijmans et al. 2009). This affects the relative rates of nucleation and crystal growth and thus determines the particle size distribution (PSD). Esposito et al. (2006) found that the use of biogenic sulfide instead of Na_2S as sulfide source for ZnS precipitation decreased the efficiency of the precipitation process both in terms of zinc effluent concentration and particle size of the precipitates.

With respect to the metal removal mechanisms involved in bioreactors, sorption onto the biofilm and carrier material (Neculita et al. 2007), complexation and precipitation with other compounds typically present in the effluent of a sulfate reducing bioreactor (phosphate, micro-nutrients, acetate and EDTA) control the fate of the metals in the system (Samaranayake et al. 2002; Esposito et al. 2006). These compete with the metal sulfide precipitation process, thus jeopardizing the success of the metal recovery.

This paper explores the influence of sulfide concentration on the settling properties of the metal precipitates as well as the contribution of other *mechanisms* to metal removal in bioreactors. To achieve this, batch experiments following metal depletion kinetics were done with a biofilm coated on polyethylene beads (obtained from an IFB reactor), synthetic wastewater containing sulfide added at different concentrations, macronutrients and heavy metals (Zn, Cu, Cd and Pb). The metals speciation and removal efficiencies were compared with the predictions of the speciation program Visual MINTEQ. X-ray diffraction (XRD) and particle size distribution (PSD) analysis were used to characterize the precipitates. In addition, X-ray absorption spectroscopy (XAS) analysis was applied to study specifically the molecular structure of Zn. These results can potentially provide a baseline for the application of inversed fluidized bed (IFB) reactors for metal recovery.

3.2 Materials and methods

3.2.1 Biomass origin

Polyethylene beads (3 mm diameter) covered with a sulfate reducing bacteria (SRB) biofilm were collected at the end of an IFB reactor operation run as described by Villa-Gómez (2011). The initial biomass content in the biofilm amounted to 3.07 mg volatile suspended solids (VSS) per gram of dry polyethylene ($VSS/g_{polyethylene}$). The initial metal concentration in the biofilm was 0.32 mg Cu/$g_{polyethylene}$, 0.52 mg Zn/$g_{polyethylene}$, 0.08 mg Pb/$g_{polyethylene}$ and 0.40 mg Cd/$g_{polyethylene}$.

3.2.2 Synthetic wastewater

The synthetic wastewater used for this study was the same to the one used in the operation of two IFB reactors by Villa-Gomez et al. (2011) and adjusted to pH 7 as optimal biological sulfate reducing conditions. It consisted of 10 mg L^{-1} Zn, Cu, Pb and Cd each as metal chlorides and macronutrients (mM): 3.74 NH$_4$Cl, 17.00 CaCl$_2$ *2H$_2$O, 3.67 KH$_2$PO$_4$, 10.14 MgSO$_4$·7H$_2$O, 0.17 FeSO4·7H$_2$O, 0.16 CuCl$_2$·2H$_2$O, 0.15 Zn Cl$_2$, 0.05 PbCl$_2$, 0.09 CdCl$_2$·2.5H$_2$O. All reagents were of analytical grade.

3.2.3 Precipitation experiments

The precipitation experiments were performed at room temperature (23 ±2 °C) in serum bottles of 117 mL containing 112 mL synthetic wastewater and 5 mL of biofilm coated on polyethylene beads or bare polyethylene beads or synthetic wastewater only. Three sets of experiments were tested at different sulfide concentrations (mM: 0, 0.62, 1.25, 2.49) as Na$_2$S and hence, different metal to sulfide molar ratios (1:0, 1:1, 1:2, 1:4). These sulfide concentrations were chosen to simulate bioreactors operating above and below the stoichiometric amounts required for metal sulfide precipitation. The synthetic wastewater contained iron (as macronutrient), which easily precipitates with sulfide, and hence, it was also considered in the calculations of the metal to sulfide molar ratio ($[\sum M^{2+}]$: The sum of Zn, Cu, Pb, Cd and Fe molar concentrations). The first precipitation experiment was done with biofilm coated on polyethylene beads (B1, B2, B3, B4), the second with bare polyethylene beads (P1, P2, P3 and P4) and the third with synthetic wastewater only (S1, S2, S3, S4). The serum bottles were stirred at 100 rpm during the whole experiment. Duplicate experiments were conducted in a non-parallel manner and the average of the duplicates is shown in the data.

Metal depletion was followed by taking samples of the liquid phase every half hour during the first 3 hours and then every hour and a half over a 9 hours period (from 0.3 mL at time zero to 1 mL at the end). This reaction time was chosen based on preliminary screening experiments (data not shown) that showed that the equilibrium was reached during this time period. Sulfide concentration was verified at time zero by taking sample of the liquid phase (0.05 mL).

3.2.4 Analytical methods

Metal measurements were done by flame spectroscopy (AAS Perkin Elmer 3110) and furnace spectroscopy (AAS Solaar MQZe GF95) after dilution, acidification with HNO_3 and filtration of the samples with a 0.45μm filter. Dissolved sulfide was determined spectrophotometrically by the colorimetric method described by Cord-Ruwisch (1985) using a Perkin Elmer Lambda20 spectrophotometer.

The precipitates of the experiments S1, S2 S3 and S4 were collected and centrifuged for further XRD and XAS analysis. XRD analyses were performed on a Bruker D8 Advance diffractometer equipped with an energy dispersion Sol-X detector with copper radiation (CuKα, λ = 0.15406 nm). The acquisition was recorded between 2° and 80°, with a 0.02° scan step and 1 s, step time. Samples were previously dried and crushed prior to XRD analysis. The sample was disposed on a Si low background sampler.

X-ray absorption near edge spectroscopy (XANES) and extended X-ray absorption fine structure (EXAFS) were performed on the DUBBLE beam line BM26A of the European Synchrotron Radiation Facility (Grenoble, France) (Borsboom et al. 1998). Zn-K-edge EXAFS spectroscopy was applied to characterize the Zn-binding sites in the precipitates of the batch experiments. Spectra were also collected from multiple reference compounds, including ZnS, Zn sorbed on apatite and $Zn_3(PO_4)_2*4H_2O$ (parahopeite and hopeite). Zn sorbed on apatite and $Zn_3(PO_4)_2*4H_2O$ (hopeite) were prepared according to Lee et al. (2005) and Pawling et al. (1999), respectively. The X-ray energy was varied from 200 eV below to 750 eV above the absorption K-edge of Zn (9659 eV). The incident intensity and transmitted intensity was measured by appropriately positioned ionization chambers. The fluorescence signal was measured with a 9 element monolithic Ge detector. During data collection, samples were maintained at a temperature of approximately 80 K using a liquid helium-flow cryostat. EXAFS spectra were processed using the X-ray absorption fine structure software package (Winterer 1997) using standard procedures (Farges et al. 2001).

Precipitates from the batch experiments S2, S3 and S4 were collected and analyzed in a particle size analyzer (Microtrac 53500) for particle size distribution (PSD) measurements after 24 hours of the termination of the experiments. Samples were also analyzed after sonication for 20 seconds at an ultrasound power of 30 KHz to differentiate the crystals from the agglomerates. The sonication was repeated until the agglomerates were completely dispersed and the values of the size distribution of the particles did not change. The mean particle size (D_{50}) of each sample before and after sonication was calculated after determining the minimum and maximum sizes contributing to the highest peak.

3.2.5 Data analysis

3.2.5.1 Determination of kinetic parameters

The kinetic parameters were determined by monitoring the depletion of the metal concentration through time. The rate constants for the precipitation of the metals studied were determined using first order equations. The kinetic modeling of the metal precipitation has been carried out using the elementary rate equation on its graphical solution (Brezonik 1994):

$$\log[M] = \log[M_0] - \frac{kt}{2.303} \tag{3}$$

Where M is the metal concentration at time "t", M_0 is the initial metal concentration and k is the metal depletion rate constant. A plot of log [M] vs. t yields a straight line for reactions following first-order kinetics. The y-intercept is log [M_0] and k is obtained from the slope, k= -2.3(slope).

3.2.5.2 Chemical equilibrium model calculations

Visual MINTEQ version 3.0 (US EPA, 1999, http://www.lwr.kth.se/English/OurSoftware/vminteq/index.html) was used to calculate speciation, solubility, and equilibrium of solid and dissolved phases for different experimental conditions to compare the predictions of the model with the obtained experimental data. Modeled data are termed M_{B1}, M_{S1}, M_{S2}, M_{S3} and M_{S4} to differentiate from experimental data. In the model, 2 mgL^{-1} of sulfide was introduced to the input data of M_{B1} (no sulfide added), since endogenous sulfide production was observed by the SRB biofilm in the batch experiments, even without any external substrate addition (data not shown).

Reactions for aqueous complexes and solid phases, including equilibrium constants, were taken from the Visual MINTEQ 3.0 thermodynamic database. During the model setup, the Davies approximation (Stumm and Morgan 1996) for activity correction was selected, oversaturated solutes were allowed to precipitate and the pH was fixed to 7.0.

Further, the program was set to output the ionic strength of the solution, the mineral identity and the amount of precipitates formed at the end of the reaction.

3.3. Results

3.3.1 Precipitation experiments

The reaction of metal and sulfide in the batch experiments was almost instantaneous in all experiments, resulting in the formation of a brownish solution with different color intensities, the highest intensity was observed at the maximum sulfide concentration added. No differences (<5%) in the metal removal rates were observed between the experiments with bare polyethylene P1, P2, P3 and P4 or with synthetic wastewater only (S1, S2, S3 and S4) (data not shown). This shows that the polyethylene itself did not induce the sorption or precipitation.

3.3.1.2 Metal depletion kinetics

Figure 3.1 shows the metal depletion rates at different sulfide concentrations for the experiments with synthetic wastewater (S1, S2, S3, S4) and biofilm (B1, B2, B3, B4). Metals precipitated and reached equilibrium concentrations within 3 hours in the following order: Cu> Pb> Cd> Zn (Figure 3.1). Moreover, an increase in the sulfide concentration clearly resulted in a more rapid Zn and Cd depletion, this behavior was less for Pb and Cu. Cu is mainly removed within the first half hour after sulfide dosing, especially for the experiments S3 and S4 (Figure 3.1o and 3.1p) and the depletion rate could not be calculated.

Zn and Cd depletion was different in batch experiments without sulfide addition (Figure 3.1a and 3.1e). In S1, 3.1 mg L^{-1} of Zn remained in the liquid phase, while only 0.1 mg L^{-1} remained in B1. For Cd, 3.3 mg L^{-1} remained in the S1 liquid phase after 9 hours while only 0.9 mg L^{-1} remained in the liquid phase for B1. The batch experiments where sulfide was added show no significant differences in metal depletion rate (except for figure 3.1f), suggesting that the biofilm did not influence metal removal in the presence of sulfide. Therefore, only S2, S3 and S4 will be considered for the remainder of this manuscript.

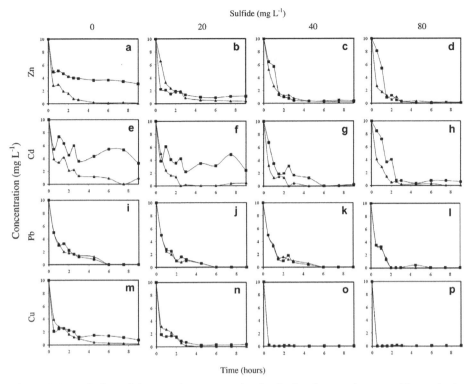

Figure 3.1 Evolution of the metal concentration in the batch experiments with synthetic wastewater (■) and biofilm on polyethylene (▲) for Zn (a-d), Cd (e-h), Pb (i-l) and Cu (m-p). Average from duplicate data.

3. 3.1.3 Kinetic parameters

Figure 3.2 shows the plots for Pb, Zn and Cd depletion (logarithmic scale) vs. time for the experiments S2, S3 and S4. The precipitation of Cu was too fast (Figure 3.1m-p) and the kinetic model described by equation 1 could not be applied to these data. Pb depletion shows a similar rate for the three experiments (Figure 3.2a), while for Cd depletion, the rate was slightly faster in experiment B4 (Figure 3.2b). The Zn depletion rate occurred in two phases: first, a characteristic linear, rapid fall during the initial 3 hours, followed by a slower removal in a second phase (Figure 3.2c).

Figure 3.2 also shows the kinetic constants for the Cd and Zn precipitation rates for the experiments S2, S3 and S4 obtained with equation 3. The kinetic constants for Pb show no significant differences for S2 and S3, reflected in Figure 3.2a, while for S4, K_{Pb} decreased to 0.300 h^{-1}. An increase of the K_{Cd} from 0.979 h^{-1} in S2 to 1.143 h in S4 was observed, while for S3, K_{Cd} was only 0.841 h^{-1}. However, experiment S3 has a relatively significant deviation ($r^2= 0.698$) compared to the other kinetic constants. This could possibly be due to a combination of several errors such as lab work or analytical error. The K_{Zn}'s were obtained for the first phase between the limits A_0, t_0 to A_3, t_3 (3 hours). K_{Zn} values show a proportional increase with the increase in sulfide concentration.

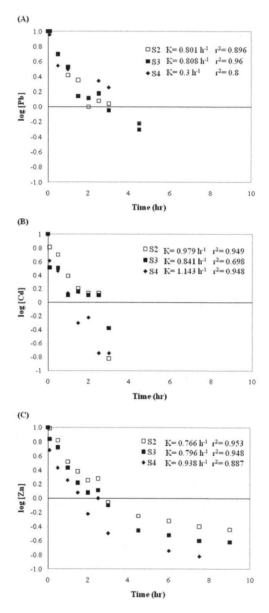

Figure 3.2 Log [A] vs. time of a) Pb, b) Cd and c) Zn concentration for experiments S2, S3 and S4.

3.3.2 Equilibrium analysis

3.3.2.1 Speciation modeling results at the equilibrium

Table 3.1 compares the modeled reactions (M) with experimental results from the batch experiments B1, S1, S2, S3 and S4. According to speciation calculations with Visual MINTEQ, metal sulfide precipitation is mainly responsible for metal depletion in batch experiments with added sulfide. Visual MINTEQ predicted at equilibrium, the formation of chalcopyrite ($CuFeS_2$), galena (PbS), greenockite (CdS) and spharelite (ZnS) for the experiments with added sulfide. In addition, the model also predicted precipitation of the phosphate supplied as macronutrient in the synthetic wastewater, mainly as hydroxylapatite $Ca_5(PO_4)_3(OH)$ (> 99.26%).

However, when sulfide is not present, the metals are removed by the biofilm (B1, Table 3.1) and by the mineral medium components (M_{S1}, Table 3.1). Cu and Pb modeled results agree with those found in the batch experiments S1 and B1 (Figure 3.1, Table 3.1). When sulfide was not added externally, Cu was found to be removed as CuOH with (M_{B1}) and without biofilm (M_{S1}) present and partly as CuS (40%) due to the endogenously produced sulfide (explained in 2.5.2) in the presence of biofilm (B1), while Pb was removed as chloropyromorphite $Pb_5(PO_4)_3(Cl)$ in both cases. In contrast, Zn and Cd results differ from the modeled results when sulfide was not added externally. In the presence of the biofilm, Zn and Cd are completely removed, while in the absence of externally supplied sulfide 69.2 and 67.5% of Zn and Cd, respectively, was removed from the liquid phase (S1, Table 3.1).

3.3.2.2 Solid phase characterization

3.3.2.2.1 XRD of the precipitates

The XRD diagrams of the precipitates are shown in Figure 3.3. Due to the poor crystallinity of the precipitates, not all the species could be identified. The identified species were brushite ($CaHPO_4 \cdot 2H_2O$) and pyromorphite ($Pb_5(PO_4)_3Cl$) for S1, while for S2, S3 and S4 only hydroxylapatite ($Ca_{10}(PO_4)_6OH$) and galena (PbS) were observed. These species were identified also in the Visual MINTEQ program, including the precipitation of the phosphate as apatite in the form of hydroxylapatite (> 99.26%), while brushite was not predicted by the Visual MINTEQ program.

67

Table 3.1 Comparison of modeled reactions (M) versus experimental results from the batch experiments B1, S1, S2, S3 and S4.

Experiment	Predicted metal removal (%)									
	M_{B1}	M_{S1}	M_{S2}	M_{S3}	M_{S4}	B1	S1	S2	S3	S4
Zn (total precipitated)	0	0	100	100	100	100	69.2	91.2	92	96.8
ZnS			100	100	100					
Cu (total precipitated)	98.9	98.9	100	100	100	100	90.7	99.7	99.7	99.7
CuS		60								
CuFeS$_2$			100	100	100					
Cu(OH)	100	40								
Pb (total precipitated)	99.98	99.98	100	100	100	100	97.16	99.02	97.5	95
PbS			100	100	100					
Pb$_5$(PO$_4$)$_3$(Cl)	100	100								
Cd (total precipitated)	0	0	100	99.97	99.99	98.9	67.5	98.5	95.8	100
CdS			100	100	100					
Other precipitates (PO$_4^{3-}$)	99.99	99.99	99.99	99.99	99.99	ND	ND	ND	ND	ND
Ca$_5$(PO$_4$)$_3$(OH)	99.23	99.23	100	100	100					

ND: Not determined

3.3.2.2.2 Zn K-edge XAS analysis

From the four metals tested, Zn has the highest solubility product and hence, it was more likely to be removed by other removal mechanisms as shown in the kinetic experiments. Therefore, XAS was used to examine the Zn speciation in the experiments. The spectra of the four experiments could fit with only 3 model compounds: ZnS, Zn sorbed on apatite and $Zn_3(PO_4)_2 \cdot 4H_2O$. Other reference compounds spectra such as $Zn(OH)_2$ were rejected in the fits of the experimental spectra. Figure 3.4 shows Zn K-edge XANES spectra of the precipitates S1, S2, S3 and S4 as well as the references compounds: ZnS and Zn sorbed on apatite. The spectra for the experiments with the highest sulfide concentrations (S3 and S4) displays high similiraties with the ZnS XANES spectra implying that Zn is mainly removed through the formation of zinc sulfide precipitate. In contrast, for the experiments with the lowest sulfide concentrations display different features compared to the ZnS XANES spectra. The comparison with the Zn sorbed apatite spectra shows that such phenomena are likely $(PO_4)_2*4H_2O$ (parahopeite and hopeite) and Zn concentrations (S3 and S4) mainly have a similar frequency as the ZnS, showing that the synthetic wastewater does have little influence on the zinc precipitation (Figure 3.5a). On the other hand, the spectra of the lowest sulfide concentrations (S1 and S2) were more (or) in phase with the oscilations of Zn sorbed on apatite and parahopeite ($Zn_3(PO_4)_2*4H_2O$ having and triclinic mineral structure) (Figure 3.5a), showing that sorption and precipitation with phosphate may contributed to the Zn depletion kinetics (Figure 3.2c).

3.5 shows the Zn K-edge

Experiment	Predicted metal removal (%)									
	M_{B1}	M_{S1}	M_{S2}	M_{S3}	M_{S4}	B1	S1	S2	S3	S4
Zn (total precipitated)	0	0	100	100	100	100	69.2	91.2	92	96.8
ZnS			100	100	100					
Cu (total precipitated)	98.9	98.9	100	100	100	100	90.7	99.7	99.7	99.7
CuS			60							
CuFeS$_2$			100	100	100					
Pb (total precipitated)	99.98	99.98	100	100	100	100	97.16	99.02	97.5	95
PbS			100	100	100					
$Pb_5(PO_4)_3(Cl)$	100	100								

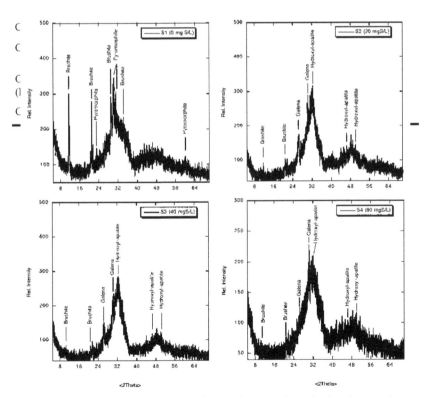

Figure 3.3 X-ray diffraction diagram of the precipitates from the batch experiments S1, S2, S3 and S4.

The Fourier transforms clearly show that at metal to sulfide ratios below 1, the nearest neighboring atom of the bound zinc ion is the oxygen atom, this in turn may be bound to phosphate: the two Zn-phosphate reference compounds and samples S1 and S2 have the major peak at near 1.5 Å (without phase shift correction), while the major peak for ZnS as well as samples S3 and S4 is at near 2 Å (without phase shift correction) (Figure 3.5b).

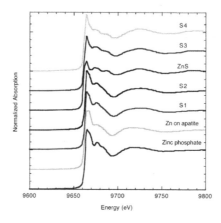

Figure 3.4 Comparison of the XANES spectra for the batch experiments S1, S2, S3, S4, and the reference compounds ZnS, $Zn_3(PO_4)_2*4H_2O$ and Zn sorbed on apatite.

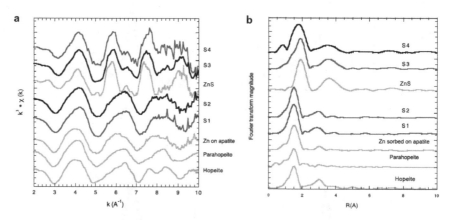

Figure 3.5 Comparison of the EXAFS spectra for the batch experiments S1, S2, S3, S4, and the reference compounds ZnS, $Zn_3(PO_4)_2*4H_2O$ (parahopeite), $Zn_3(PO_4)_2*4H_2O$ (hopeite) and Zn sorbed on apatite. a) Zn K edge k^3-weighted χ (k) curves and b) Fourier transform EXAFS corresponding radial structure functions (not corrected for phase shifts).

3.3.2.2.3 PSD of the precipitates

Figure 3.6 shows the effect of the sulfide concentration on the average particle size of the produced solids. The PSD's determined before sonication and hence, considering the agglomeration of the particles, were reasonably close in the region, but differed in decreasing order of diffraction volume. The same behavior was found after sonication, but the range of the region was different (0 - 70 μm). The volume fraction of the particle size was greatly reduced after sonication and the fractions of some large particles were also decreased. The D_{50} for the particles before sonication (considering agglomeration) was in the range of 14.8-15.2 μm, and decreased to 8.1-10.0 μm after sonication. The D_{50} value obtained for the particles after sonication shows a slight decrease with the increment of the sulfide concentration.

3.4 Discussion

3.4.1 Metal depletion kinetic parameters

This study shows that the metal depletion kinetics and removal mechanisms highly depend on the sulfide concentration (Figure 3.1, 3.4 and 3.5 and Table 3.1), while the growth and settling of the precipitates for potential recovery is more associated with an agglomeration phenomenon (Figure 3.6). The depletion rate of the metals when sulfide was added followed the same precipitation order according to the solubility product of metal sulfides (Sampaio et al. 2009). The sequential precipitation in multimetal systems has been described previously by the manipulation of the sulfide concentration and pH (Sahinkaya et al. 2009; Sampaio et al. 2009; Veeken et al. 2003). Sampaio et al. (2009) achieved the selective recovery of Cu over Zn due to the difference in solubility product of CuS and ZnS. However, these studies do not acknowledge that the extremely small particle sizes produced by the metal sulfide precipitation processes do not lead themselves to solid liquid separation (Lewis 2010). Metal precipitation, as a chemical reaction, can occur in seconds forming colloids at nanometer scale. However, particle growth and hence, the observed metal depletion occurs at a slower rate (Peters et al. 1984; Luther and Rickard 2005). This differentiates the metal precipitation from metal depletion rates due to settling for real potential recovery. In the present study, the second argument was studied, and hence, a sequential metal depletion could be observed (Figure 3.1). The potentiality for metal recovery has already been demonstrated by Mishra and Das (1992). They observed that the precipitation kinetics for cobalt sulfide show an induction period of a few minutes during which ZnS is precipitated. Based on this, a two stages precipitation was used obtaining 90% of Zn separated from cobalt.

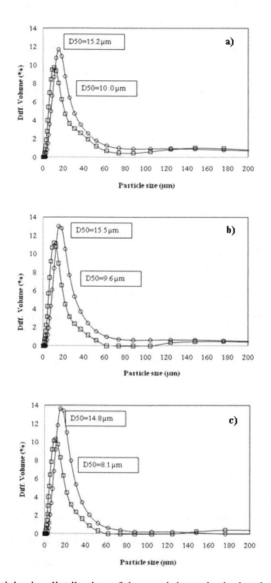

Figure 3.6 Particle size distribution of the precipitates in the batch experiments a) S2, b) S3 and c) S4 before ○ and after sonication □.

The different K values from the plots of Pb, Cd and Zn depletion for the different sulfide concentrations support the first-order kinetic model (Figure 3.2). Several authors have found that the kinetics of the metal sulfide reactions is well described by a first order equation (Bryson and Bijsterveld 1991; Mishra and Das 1992; Lewis and Swartbooi 2006). For the metals Mn, Ni, Co and Zn, they showed a better data fit with a standard deviation close to 1, in contrast with this study (Figure 3.2). One reason is the presence of macronutrients that interfere in the metal sulfide precipitation process and causes competing precipitation reactions (Brezonik 1994). Another reason is the difference in solubility product of the metals of study. Mn, Ni, Co and Zn have the highest solubility product over the range of metals and hence, slower reaction rates that can be better described. In this study, Zn and Cd had a better fit (Figure 3.1a-h) compared to Pb. The latter has a lower solubility product and thus, faster precipitation kinetics (Figure 3.1i-p), whereas Cu could not be described at the sampling time interval chosen in this study. CuS precipitation has been reported to occur instantaneously (Sampaio et al. 2009; Sahinkaya et al. 2009; Lewis and van Hille 2006; van Hille et al. 2005). Sahinkaya et al. (2009) observed that the Cu sulfide precipitation occurred within 30-50 min. in batch experiments with biogenic sulfide at metal to sulfide ratios of 1.4:1, 0.74:1 and 0.47:1 (229-342 mg L^{-1} of sulfide and 50-100 mg L^{-1} of Cu).

3.4.2 Metal removal in the absence of sulfide

This study also shows that metal sulfide precipitation can only be achieved when the sulfide concentration is maintained above stoichiometry and hence, metal sulfide precipitation is favoured. The differences between the experiments with and without biofilm when sulfide was not externally added (Figure 3.1) clearly show that Zn and Cd are partly removed via the biofilm, either through an ion exchange mechanism on the surface of the biosorbent (biofilm) or surface precipitation of metal hydroxide/sulfide species (van Hullebusch et al. 2003). A comparison between batch growth tests (bioprecipitation) and dead biomass batch tests (biosorption) with SRB by Pagnanelli et al. (2010) showed that Cd was mainly removed by a biosorption mechanism (77%) due to metabolism-independent binding properties of the SRB cell wall surface. Nevertheless, this sorption becomes less strong with the gradual occupancy of the active sites. This was confirmed when the experiment was repeated using the same biofilm from previous experiments (data not shown): metal depletion rate gradually behaved similar to the experiments without biofilm.

Another metal removal mechanism is the precipitation with components of the synthetic wastewater. The presence of hydroxides and phosphates, contained in the synthetic wastewater, caused the precipitation of 40% of the Cu and 100% of the Pb at low sulfide concentrations according to Visual MINTEQ prediction (Table 3.1), the hydroxide precipitation of metals has been reported in passive sulfate reducing reactors operating at a pH around 7 during the first stages of operation, when sulfate reduction is not yet well established and hence, low sulfide production occurs (Samaranayake et al. 2002; Neculita et al. 2008). With respect to phosphate precipitation, several studies have shown the potential of phosphates for the immobilization of divalent heavy metals like Zn, Pb, Cu, and Cd from wastewaters, solid wastes and contaminated soils (Geysen et al. 2004). However, this removal mechanism has not been taken into account in sulfate reducing reactors. Bartacek et al. (2008) showed that cobalt depletion might be due to phosphates and carbonates precipitation in a study on the influence of cobalt speciation on the toxicity of cobalt to methylotrophic methanogenesis in anaerobic granular sludge.

3.4.2.1 Zn removal mechanism

In this study, the analysis of the Zn coordination with EXAFS analysis confirmed that apatite and Zn phosphates precipitates (parahopeite) may contributed to the Zn removal not only when sulfide was not externally added, but also in the presence of low concentrations of sulfide. This removal mechanism was not observed on XRD analysis or the Visual MINTEQ program. XRD analysis is effectively used when the precipitates are crystallized such as metal sulfides in nature. However, metal sulfides prepared in the laboratory have shown to be poorly crystallized limiting its analysis with XRD or other techniques such as scanning electron microscopy (Neculita et al. 2007). On the other hand, equilibrium calculations done with the geochemical program cannot predict solution concentrations of ions in contact with solids formed in the system. Hence, a removal of Zn with another compound formed in the system such as apatite could not be predicted using Visual MINTEQ or observed with XRD. Chen et al. (1997) explained that the interaction of apatite with heavy metals may form relatively insoluble metal phosphates and/or result in the sorption of heavy metals on apatite, thus significantly reducing aqueous metal concentrations. A sorption mechanism of Ca^{2+} ion exchange with a divalent metal such as Zn^{2+} on apatite ($Ca_5(PO_4)_3^+$ (Cl^-, OH^-, F^-, Br^-)) has been described as responsible for the formation of the precipitates. Aklil et al. (2004) showed that hydroxyapatite functions as an adsorbent capable to remove several toxic metals in batch experiments with calcinated phosphate. The absorption capacities they obtained at pH 5 were 85.6, 29.8 and 20.6 mg/g for Pb, Cu and Zn, respectively.

3.4.3 Particle size: crystallization and agglomeration for potential metal recovery

The recovery of metals in bioreactors highly depends on the size of the particles formed. Large particles can be produced if the supersaturation is optimum to promote crystal growth, and the residence time of the crystals is long enough to promote agglomeration (Mersmann 1999). In this study it was observed that the sulfide concentration, and hence, supersaturation, barely varied the size of the metal precipitates, while agglomeration was more relevant for the increase of the particles (Figure 3.6). These results agree with other authors that have described metal sulfide particle growth characteristics (Al-Tarazi et al. 2004; Esposito et al. 2006; Sampaio et al. 2009). They have observed that the degree of supersaturation does not necessarily control the rate of particle growth since, due to the fast reaction kinetics of metal sulfides, primary nucleation predominates instead of crystallization and hence, only small particles with poor settling properties are formed. Consequently, agglomeration has been shown to be the predominant mechanism to obtain bigger particles in the precipitation of metal sulfides.

The use of bioreactors might be beneficial to promote agglomeration of metal sulfides due to the presence of substances that contribute to this mechanism. Peters et al. (1984) found that, for ZnS precipitated at pH 8, the complexing agents (ammonia, EDTA and 18-Crown-6 ether) reduced the nucleation rate, but promoted the aggregation rate. However, agglomerates are often relatively fragile and hence, the hydrodynamics might play a crucial role on the shape of the agglomerates obtained. Sampaio et al. (2009) observed that the PSD of CuS particles increased if allowed to settle (from 36 to 180 μm) and if the same sample was vigorously stirred, then the PSD decreased to below 3 μm. This study reports the agglomeration of the precipitates upon termination of the experiment (Figure 3.6) and with substances commonly present in bioreactors (e.g. macronutrients). However, the agglomeration process needs to be study in bioreactors in order to determine the effect of the hydrodynamics of the system on the PSD of the metal precipitates.

3.5 Conclusions

- Metal sulfide precipitation can only be achieved when the sulfide concentration is maintained above stoichoimetry and hence, metal sulfide precipitation is favoured.

- Zn and Cd, which are the metals with the highest solubility product in this study, were removed by metal removal mechanisms other than metal sulfide precipitation. These include biosorption, precipitation with macronutrients as phosphate and sorption onto their precipitates, e.g. apatite.

- EXAFS analysis of Zn coordination confirmed that apatite and Zn phosphate precipitates contributed to Zn removal.

- Agglomeration is the most important mechanism for particle growth and settling of the metal precipitates. However, hydrodynamics affect the agglomeration process. This needs to be studied in more detail in continuously operated bioreactors.

References

Aklil A, Mouflih M, Sebti S (2004) Removal of heavy metal ions from water by using calcined phosphate as a new adsorbent. Journal of Hazardous Materials 112 (3):183-190

Al-Tarazi M, Heesink ABM, Azzam MOJ, Yahya SA, Versteeg GF (2004) Crystallization kinetics of ZnS precipitation; an experimental study using the mixed-suspension-mixed-product-removal (MSMPR) method. Crystal Research and Technology 39 (8):675-685. doi:10.1002/crat.200310238

Bartacek J, Fermoso F, Baldó-Urrutia A, van Hullebusch E, Lens P (2008) Cobalt toxicity in anaerobic granular sludge: influence of chemical speciation. Journal of Industrial Microbiology & Biotechnology 35 (11):1465-1474. doi:10.1007/s10295-008-0448-0

Bijmans MFM, van Helvoort P-J, Buisman CJN, Lens PNL (2009) Effect of the sulfide concentration on zinc bio-precipitation in a single stage sulfidogenic bioreactor at pH 5.5. Separation and Purification Technology 69 (3):243-248

Borsboom M, Bras W, Cerjak I, Detollenaere D, Glastra van Loon D, Goedtkindt P, Konijnenburg M, Lassing P, Levine YK, Munneke B, Oversluizen M, van Tol R, Vlieg E (1998) The Dutch-Belgian beamline at the ESRF. Journal of Synchrotron Radiation 5 (3):518-520. doi:doi:10.1107/S0909049597013484

Brezonik PL (1994) Chemical kinetics and process dynamics in aquatic systems. Lewis Publishers, Boca Raton, Florida

Bryson AW, Bijsterveld CH (1991) Kinetics of the precipitation of manganese and cobalt sulphides in the purification of a manganese sulphate electrolyte. Hydrometallurgy 27 (1):75-84

Cord-Ruwisch R (1985) A quick method for the determination of dissolved and precipitated sulfides in cultures of sulfate-reducing bacteria. Journal of Microbiological Methods 4 (1):33-36

Chen, Wright JV, Conca JL, Peurrung LM (1997) Effects of pH on Heavy Metal Sorption on Mineral Apatite. Environmental Science & Technology 31 (3):624-631. doi:10.1021/es950882f

Esposito G, Veeken A, Weijma J, Lens PNL (2006) Use of biogenic sulfide for ZnS precipitation. Separation and Purification Technology 51 (1):31-39

Farges F, Brown GE, Petit P-E, Munoz M (2001) Transition elements in water-bearing silicate glasses/melts. part I. a high-resolution and anharmonic analysis of Ni coordination environments in crystals, glasses, and melts. Geochimica et Cosmochimica Acta 65 (10):1665-1678

Fu F, Wang Q (2011) Removal of heavy metal ions from wastewaters: A review. Journal of Environmental Management 92 (3):407-418. doi:10.1016/j.jenvman.2010.11.011

Gallegos-Garcia M, Celis LB, Rangel-Méndez R, Razo-Flores E (2009) Precipitation and recovery of metal sulfides from metal containing acidic wastewater in a sulfidogenic down-flow fluidized bed reactor. Biotechnology and Bioengineering 102 (1):91-99

Geysen D, Imbrechts K, Vandecasteele C, Jaspers M, Wauters G (2004) Immobilization of lead and zinc in scrubber residues from MSW combustion using soluble phosphates. Waste Management 24 (5):471-481

Kaksonen AH, Puhakka JA (2007) Sulfate Reduction Based Bioprocesses for the Treatment of Acid Mine Drainage and the Recovery of Metals. Engineering in Life Sciences 7 (6):541-564

Lee YJ, Elzinga EJ, Reeder RJ (2005) Sorption Mechanisms of Zinc on Hydroxyapatite: Systematic Uptake Studies and EXAFS Spectroscopy Analysis. Environmental Science & Technology 39 (11):4042-4048. doi:10.1021/es048593r

Lewis A, Swartbooi A (2006) Factors Affecting Metal Removal in Mixed Sulfide Precipitation. Chemical Engineering & Technology 29 (2):277-280. doi:10.1002/ceat.200500365

Lewis A, van Hille R (2006) An exploration into the sulphide precipitation method and its effect on metal sulphide removal. Hydrometallurgy 81 (3-4):197-204

Lewis AE (2010) Review of metal sulphide precipitation. Hydrometallurgy 104 (2):222-234

Luther G, Rickard D (2005) Metal Sulfide Cluster Complexes and their Biogeochemical Importance in the Environment. Journal of Nanoparticle Research 7 (4):389-407. doi:10.1007/s11051-005-4272-4

Mersmann A (1999) Crystallization and precipitation. Chemical Engineering and Processing 38 (4-6):345-353

Mishra PK, Das RP (1992) Kinetics of zinc and cobalt sulphide precipitation and its application in hydrometallurgical separation. Hydrometallurgy 28 (3):373-379

Neculita C-M, Zagury GJ, Bussiere B (2007) Passive Treatment of Acid Mine Drainage in Bioreactors using Sulfate-Reducing Bacteria: Critical Review and Research Needs. J Environ Qual 36 (1):1-16. doi:10.2134/jeq2006.0066

Neculita C-M, Zagury GJ, Bussière B (2008) Effectiveness of sulfate-reducing passive bioreactors for treating highly contaminated acid mine drainage: II. Metal removal mechanisms and potential mobility. Applied Geochemistry 23 (12):3545-3560

Pagnanelli F, Cruz Viggi C, Toro L (2010) Isolation and quantification of cadmium removal mechanisms in batch reactors inoculated by sulphate reducing bacteria: Biosorption versus bioprecipitation. Bioresource Technology 101 (9):2981-2987

Pawlig O, Trettin R (1999) Synthesis and characterization of [alpha]-hopeite, Zn3(PO4)2·4H2O. Materials Research Bulletin 34 (12-13):1959-1966. doi:10.1016/s0025-5408(99)00206-8

Peters RW, Chang T-K, Ku Y (1984) Heavy metal crystallization kinetics in an MSMPR crystallizer employing sulfide precipitation. Journal Name: AIChE Symp Ser; (United States); Journal Volume: 80:240:Medium: X; Size: Pages: 55-75

Sahinkaya E, Gungor M, Bayrakdar A, Yucesoy Z, Uyanik S (2009) Separate recovery of copper and zinc from acid mine drainage using biogenic sulfide. Journal of Hazardous Materials 171 (1-3):901-906

Samaranayake R, Singhal N, Lewis G, Hyland M (2002) Kinetics of biochemically driven metal precipitation in synthetic landfill leachate. Remediation Journal 13 (1):137-150

Sampaio RMM, Timmers RA, Xu Y, Keesman KJ, Lens PNL (2009) Selective precipitation of Cu from Zn in a pS controlled continuously stirred tank reactor. Journal of Hazardous Materials 165 (1-3):256-265

Stumm W, Morgan JJ (1996) Aquatic chemistry : chemical equilibria and rates in natural waters. Environmental science and technology 3rd ed edn., New York

van Hille RP, A. Peterson K, Lewis AE (2005) Copper sulphide precipitation in a fluidised bed reactor. Chemical Engineering Science 60 (10):2571-2578

van Hullebusch E, Zandvoort M, Lens P (2003) Metal immobilisation by biofilms: Mechanisms and analytical tools. Reviews in Environmental Science and Biotechnology 2 (1):9-33

Veeken AHM, de Vries S, van der Mark A, Rulkens WH (2003) Selective Precipitation of Heavy Metals as Controlled by a Sulfide-Selective Electrode. Separation Science and Technology 38 (1):1 - 19

Villa-Gomez D, Ababneh H, Papirio S, Rousseau DPL, Lens PNL (2011) Effect of sulfide concentration on the location of the metal precipitates in inversed fluidized bed reactors. Journal of Hazardous Materials 192 (1):200-207. doi:10.1016/j.jhazmat.2011.05.002

Winterer M (1997) XAFS - A Data Analysis Program for Materials Science. J Phys IV France 7 (C2):C2-243-C242-244

CHAPTER

4

EFFECT OF
HYDRAULIC RETENTION TIME
ON THE SULFATE REDUCTION
PROCESS
AND
METAL PRECIPITATES
CHARACTERISTICS

Abstract

The change in size of the metal (Cu, Zn, Pb and Cd) sulfide precipitates as a function of the hydraulic retention time (HRT) was studied in two sulfate reducing inversed fluidized bed (IFB) reactors operating at different chemical oxygen demand concentrations to produce high and low sulfide concentrations. The change in HRT from 24 to 9 h affected the size of the metal sulfide precipitates due to the change in contact time allowing aggregation and growth of the metal sulfides. The further HRT decrease to 4.5 h affected the sulfate reducing activity for sulfide production and hence, the supersaturation level. Moreover, the sulfide concentration at an HRT of 4.5 h was below the metal to sulfide stroichiometric ratio, resulting in free metal toxicity and subsequent process failure as well as metal precipitation other than metal sulfides. The lowering of the HRT decreased the size of the SRB population and their activity in the biofilm, while this effect was less strong in the biomass suspended in the reactor liquid.

This chapter was submitted for publication as:

Villa-Gomez D. K., Enright A.M., Listya E., Buttice A., Kramer, H. and Lens P.N.L. (2013) Effect of hydraulic retention time on metal precipitation in sulfate reducing inverse fluidized bed reactors. Submitted to Journal of Chemical Technology and Biotechnology.

4.1 Introduction

Metal-containing wastewaters represent an environmental and human health problem, but also a potential resource of valuable metals when these can be recovered (Lens et al. 2002). Several technologies have been applied to treat this type of wastewaters including chemical precipitation, ion exchange, adsorption, coagulation-flocculation and membrane filtration (Fu and Wang 2011). Not all these technologies are, suitable for metal recovery and may lead to high disposal expenses (Tabak et al. 2003; Esposito et al. 2006). Metal sulfide precipitation enables the recovery of metals from these wastewaters and thus improves the economic value of the metal removal process (Kaksonen et al. 2003).

Biologically produced (biogenic) sulfide by sulfate reducing bacteria (SRB) has become an alternative for sulfide generation since these wastewaters in most cases also contain sulfate (Boonstra et al. 1999). Simultaneous biogenic sulfide production and metal sulfide precipitation in a bioreactor simplifies the process design, reduces investment costs, and avoids transport of toxic sulfide to a precipitator reactor (Bijmans et al. 2009). In this context, the inverse fluidized-bed reactor (IFB) is an attractive configuration to allow recovery of metal sulfides at the bottom of the reactor, separated from the floating SRB biomass (Villa-Gomez et al. 2011a; Gallegos-Garcia et al. 2009). Metal recovery, however, can only be possible if the particle size of the metal sulfide precipitates is increased, allowing the particles to settle. The size of the precipitates depends on the kinetics of nucleation, crystal growth and agglomeration (Al-Tarazi et al. 2004). Large particles can be produced if the supersaturation is optimum to promote crystal growth, and the residence time of the crystals is long enough to promote agglomeration (Mersmann 1999). Several studies have correlated the size of the metal sulfide precipitates with the change in operational parameters such as pH, sulfide concentration (Sahinkaya et al. 2009; Sampaio et al. 2009; Veeken et al. 2003; Bijmans et al. 2009; Al-Tarazi et al. 2005b) and reactor configuration (Al-Tarazi et al. 2005a). These studies show that nucleation outcompetes the crystallization mechanism due to the low solubility of metal sulfides, which results in high levels of supersaturation. High supersaturation leads to the formation of small precipitates known as "fines" with poor settling properties (Lewis 2010). Several authors have reported that the increase in particle size is mainly a consequence of aggregation or agglomeration (larger aggregates) of these fines (Al-Tarazi et al. 2005a; Al-Tarazi et al. 2005b). Agglomerates are, however, relatively fragile and so, the hydrodynamics in the bioreactor might play a crucial role in the size of the obtained precipitates (Frédéric et al. 2009).

The hydraulic retention time (HRT) is a hydrodynamic parameter that has, up to now, only been studied with a focus on sulfate reduction and metal removal efficiencies (Kaksonen et al. 2004; Kaksonen et al. 2006; Neculita et al. 2008), but its effect on metal precipitate characteristics such as particle size for settling has not yet been addressed. The variation of the HRT can also lead to process failure as a consequence of biomass washout or metal toxicity due to shock loadings. Information on the microbial community structure is thus useful to explain the changes in bioreactor performance and metal recovery upon variations in the HRT. Recently, several authors have indeed shown the potential of using quantitative information about the microbial population structure in diagnosing problems of anaerobic bioreactor operation (Bialek et al. 2011; Siggins et al. 2011). The objective of this study was, therefore, to analyze the differences in the metal precipitation characteristics and the SRB activity and community upon changes in the HRT of sulfate reducing IFB reactors.

4.2 Materials and methods

4.2.1 IFB reactor set-up
The experiments were conducted at 25 (± 3) °C in two IFB reactors with the same configuration as described by Villa-Gomez et al. (2011a). The IFB reactors consisted of a 2.5 L conical bottom column with 0.05 m diameter and 1 m height. The carrier-bed expansion was maintained at 30% of the working volume of the IFB reactor column. The reactors were inoculated with 200 mL of polyethylene beads (3 mm diameter) covered with a SRB biofilm developed from two IFB reactors operating for 188 days described previously by Villa-Gomez et al. (Villa-Gomez et al. 2011a).

4.2.2 Synthetic wastewater
The synthetic wastewater used during reactor operation, batch experiments and metal precipitation experiments contained (mg/L): KH_2PO_4 500, NH_4Cl 200, $CaCl_2 \cdot 2H_2O$ 2500, $FeSO_4 \cdot 7H_2O$ 50, $MgSO_4 \cdot 7H_2O$ 2500 and micronutrients as described by Zehnder et al. (1980). Cu, Pb, Cd and Zn were added after the start-up period in the reactors. The yeast extract and the micronutrients were omitted from the synthetic wastewater when

metals were added in the reactors and in the metal precipitation experiments. The pH of the medium was adjusted to 7.0 ($\pm$0.5) with NaOH. Lactate was used as electron donor and sulfate (SO_4^{2-}) was supplemented as Na_2SO_4 to maintain a chemical oxygen demand (COD)/SO_4^{2-} ratio of 1 (g/g). All reagents were of analytical grade.

4.2.3 Experimental design

Figure 4.1 gives a schematic representation of the experimental design. Reactor 1 (R1) and reactor 2 (R2) were run during the first 26 days at a COD concentration of 2 g/L. After this start-up period, R1 was maintained at the same operational conditions, whereas R2 was operated at an influent COD concentration of 3 g/L but maintaining the COD/SO_4^{2-} ratio of 1. On day 59, Cu, Pb, Cd and Zn were added as chloride salts to the influent of both reactors at a concentration of 100 mg/L each. The HRT in both reactors was decreased from 24 h to 9 h on day 117 and to 4.5 on day 164. During the reactor operation at an HRT of 4.5 h, the reactors failed in terms of sulfate and COD removal efficiency and therefore, on day 173, they were operated without metals for seven days to allow the recovery of the process. After this, metals were re-supplied to the synthetic wastewater.

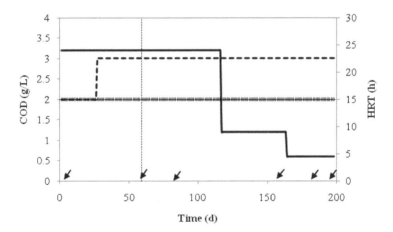

Figure 4.1 Schematic representation of the changes in COD concentration and HRT (solid line) in R1 (+) and R2 (-). The vertical dashed line indicates the starting of Cu, Zn, Cd and Pb addition. Arrows indicate the biomass sampling times for molecular phylogenetic analysis.

4.2.4 Metal precipitation experiments

The metal depletion rate in the liquid phase due to metal sulfide formation and precipitation was monitored in batch experiments to underpin the metal precipitation results in the bioreactors. The experiments were performed in triplicate at room temperature (25 ± 3 °C) in serum bottles of 117 mL with 112 mL of synthetic wastewater, containing initially 100 mg/L each of Cu, Pb, Cd and Zn and 300 mg/L of sulfide as Na_2S. This sulfide concentration was chosen as an average of the sulfide concentration produced in the IFB reactors. Although chemically produced sulfide was used instead of biogenic sulfide, the results represented adequately the settling properties of the metal sulfide precipitates affecting the metal removal efficiencies obtained in the IFB reactors. The serum bottles were stirred at 100 rpm during the experiment. Samples of the liquor were taken every half hour during 4.5 h. To differentiate metal sulfide formation (precipitation) from metal sulfide settling, two different sampling methods were tested. Samples of the liquor of a first set where filtered through 0.45 µm cellulose syringe filters and then acidified, while samples of a second set where first acidified and then filtered. All samples were analyzed for their metal concentration.

4.2.5 Sulfate reducing activity tests

Polyethylene beads covered with biofilm and bioreactor liquid samples were taken for sulfate reducing activity (SRA) tests before and after metal supply and upon termination of the IFB reactor operation. The SRA experiments were performed as described by Villa-Gomez et al. (Villa-Gomez et al. 2011a) with the synthetic medium used in this study and 2 g/L of COD (as lactate) and SO_4^{2-} at a COD/SO_4^{2-} ratio of 1. SRA tests with the suspended biomass from the bioreactor liquid contained 50 mL of the bioreactor liquid from the reactors and 64 mL of synthetic wastewater with COD and sulfate as in the SRA biofilm experiments. Since the bioreactor liquor samples already contained lactate, the COD concentration in the serum bottles was adjusted to 1 gCOD/L at the beginning of the experiments. Prior to inoculation, the bioreactor liquor sample was bubbled with N_2 for 15 minutes to remove previously produced sulfide. The polyethylene beads covered with biofilm and the bioreactor liquor samples were stored at 4 °C. Therefore, the experiments started after the activation of the biomass by its incubation under the same experimental conditions for 72 h. The bottles with biofilm were drained to remove the old media and then re-suspended in new media. All experiments were done in duplicate with a control without COD addition.

4.2.6 DNA extraction

Polyethylene beads covered with biofilm and bioreactor liquid samples were collected from both IFB reactors prior to each change in operational condition (Figure 4.1) for phylogenetic analysis of the bacterial community, specifically SRB. Total DNA was extracted from both biofilm and bioreactor liquid samples using an automated nucleic acid extractor (Magtration 12 GC, PSS Co., Chiba, Japan) as follows: polyethylene beads covered in biofilm were first suspended in 150 µl of PCR grade H_2O and sonicated for 30 seconds, ensuring the detachment of the biofilm. 100 µl of the solution was then used in the automated extraction process; 100 µl of the liquor samples were used directly in the automated extractor. Each extraction was performed in duplicate and the extracted DNA was eluted in PCR grade H_2O and stored at - 4°C.

4.2.7 Real-time PCR analysis

One bacterial primer and probe set was used to quantify the total bacterial population (Yu et al. 2005; Lee et al. 2009) while the primers and probe sets used for the SRB population were DSR4R (5′-GTGTAGCAGTTACCGCA-3′) and DSRp2060F (5′-GC clamp-CAACATCGTYCAYACCCAGGG-3′) as described by Foti et al. (Foti et al. 2007). A pure culture of *Desulfovibrio longus* 6739[T] (Magot et al. 1992), grown up in *Desulfovibrio* medium no. 63 (DSMZ) was used to generate the SRB standard curve. All DNA samples were analyzed in duplicate.

Reaction mixtures to target the total bacterial population were prepared using the LightCycler TaqMan Master kit (Roche): 8 µl of PCR-grade water, 1 µl of the probe (final conc. 200nM), 1 µl of each primer (final conc. 500 nM), 4 µl of 5x reaction solution and 5 µl of DNA template. Amplification was carried out using a two-step thermal cycling protocol consisting of pre-denaturation at 94 °C for 10 min, followed by 40 cycles at 94 °C for 10 s and 60 °C for 30 s. The SRB reaction mixture was prepared using the LightCycler 480 SYBR Green I Master kit (Roche): 3 µl of PCR-grade water, 10 µl of SYBR green reaction solution (final conc. 200nM), 1 µl of each primer (final conc. 500 nM), and 5 µl of DNA template. The amplification consisted of 45 cycles, with 1 cycle of denaturation (95°C for 40 s), annealing (55°C for 40 s), and elongation

(72°C for 1 min). Quantitative standard curves were constructed using the standard plasmids containing the full-length 16S rRNA gene sequences (Lee et al. 2009; Yu et al. 2005) and the dsrB gene subunit from the representative strains of the target bacterial and SRB groups since it encodes key enzymes of the SRB energy metabolism (Wagner et al. 2005). Results are presented in copies per µl of sample (liquid phase or biofilm support volume).

4.2.8 Chemical analysis

Biofilm detachment for volatile suspended solids (VSS) analysis was carried out as described by Villa-Gomez et al. (2011a). After this, VSS in both biofilm and bioreactor liquid samples was determined according to standard methods (APHA 2005). COD was determined by the close reflux method (APHA 2005). Sulfate and metals determination followed the same procedure described by Villa-Gomez et al. (2011a). Sulfide was determined spectrophotometrically by the colorimetric method described by Cord-Ruwisch (1985) using a spectrophotometer (Perkin Elmer Lambda20, Groningen, The Netherlands).

The metal precipitates were recovered from the bottom of the reactors at different days during the different operational periods. The samples were collected and analyzed in a particle size analyzer (Microtrac 53500) for particle size distribution (PSD) measurements after sonication for 30 seconds at an ultrasound power of 30 KHz. This technique disrupts flocs and unstable agglomerates (Manual, Microtrac S3500) in order to enable quantification of metal sulfide precipitates at all ranges, including the small particles that tend to aggregate as well as the stable agglomerates. The sonication was repeated until the values of the size distribution of the particles did not change (about 3 times) and taking care to avoid over-sonication that may dissolve fine particles or increase the particle surface energy leading to re-agglomeration (Manual, Microtrac S3500). All samples showed around five peaks over the size range, and hence, the results are presented in percentage of particles at the different size ranges.

Visual MINTEQ version 3.0 (US EPA, 1999) was used to calculate speciation, saturation indices and theoretical precipitates in the IFB reactor at the sulfide concentrations observed during the IFB reactor operation and in the synthetic wastewater prior to enter the system. The procedure is explained in detail by Villa-Gomez et al. (Villa-Gomez et al. 2012a).

4.3 Results

4.3.1 IFB reactor performance

The IFB reactors were operated at the start up under the same conditions during 27 days, and after this, the COD in the influent of R2 was increased to produce different sulfide concentrations to study its effect on metal precipitation. After the change in the COD influent concentration in R2, the COD removal efficiency was 52.3% for R1 and 40.4% for R2 and the sulfate removal efficiency was 44.6% and 35.2%, respectively. The sulfide concentration in R2 increased considerably after the change in the COD influent concentration to 3 g/L, reaching values around 850 mg/L (Figure 4.2) that were kept for 4 days and then started to decrease until it was maintained at a mean sulfide concentration of 567.9 mg/L (Table 4.1), while the sulfide concentration in R2 was stable around 441.2 mg/L prior to metal addition.

Table 4.1 Sulfide concentration, COD and sulfate removal efficiency in both reactors during the different operational conditions (mean ± standard deviation).

			Metal addition		
			HRT 24	HRT 9 Hr	HRT 4.5*
R1	COD (%)	52.3 ± 10.6	41.4 ± 12.1	25.56 ± 5	46.5 ± 11.9
	SO_4^{2-} (%)	35.2 ± 10.2	36.82 ± 13.3	34.55 ± 10.5	9.6 ± 12.9
	Sulfide (mg/L)	441.2 ± 92	290 ± 41.1	301.87 ± 57.7	54.9 ± 87.6
R2	COD (%)	40.4 ± 10.2	39.5 ± 9.5	18.4 ± 4.4	12.2 ± 2.2
	SO_4^{2-} (%)	44.6 ± 6.5	44.3 ± 9.9	33.8 ± 10.7	14.7 ± 15.3
	Sulfide (mg/L)	567.9 ± 41.4	540 ± 53.9	379 ± 96	110.2 ± 144.9

*Results obtained after re-starting metal addition.

After metals were added to the influent, the sulfide concentration increased, probably due to an increase in the SRB activity once the sulfide reacted with metals and thus alleviating SRB sulfide inhibition (Ferris et al. 1989). This lasted 4 days and then, the sulfide concentration decreased (Figure 4.2) until it established at around 290 mg/L (Table 4.1). The COD removal efficiency also decreased in R1 (41.4 %), while in R2, the values remained similar to the values obtained prior to metal addition (Table 4.1). During the reactor operation at an HRT of 9 h, the COD removal efficiency decreased considerably in both reactors to 25.6% and 18.4%, respectively, while the sulfate removal efficiency and the mean sulfide concentration decreased only in R2 (Table 4.1,

Figure 4.2). The further decrease of the HRT to 4.5 h strongly affected the sulfate reduction efficiency in the two IFB reactors (Figure 4.2). During this operation mode, the interruption of the metal feeding induced a partial recovery of the COD and sulfate removal efficiency in both reactors (data not shown). After re-starting the metal addition to the influent, however, both the sulfate and COD removal efficiency dropped again to values below 14.7%, except for the COD removal efficiency (46.5%) in R1 (Table 4.1). The decrease in HRT from 24 to 9 and latter to 4.5 h produced stronger fluctuations in the sulfide concentration (Figure 4.2).

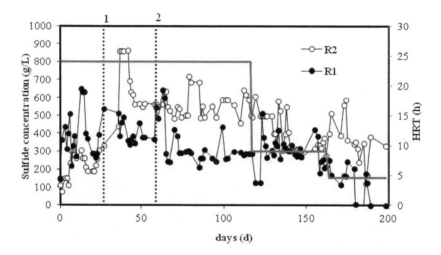

Figure 4.2 Sulfide concentration profile during the changes in operational conditions in the IFB reactors. Dashed line 1 and 2 indicate the change in COD influent concentration in R2 and the start of metal addition, respectively.

4.3.2 Metal removal efficiency in the IFB reactors

Figure 4.3 shows the metal removal efficiencies for Cu, Zn and Cd at the HRTs of 24, 9 and 4.5 h in R1 and R2. The Pb removal efficiencies are not reported since the measured values in the influent were very low already (1 mg/L) despite that no precipitates were visible in the influent tank. Pb concentrations were also below 1 mg/L in the effluent at all HRTs applied.

Up to 99% of the metals were removed from the influent at an HRTs of 24 h in both reactors, while at an HRT of 9 h, the Cu removal efficiency remained over 98% on

average and the Zn and Cd removal efficiency fluctuated around values from 34.8 to
99.0% in R1 and from 39.8 to 99.0% in R2 (Figure 4.3).

At the HRT of 4.5 h, metal removal efficiencies exceeded 50%, except for Cd with no
removal on day 182 (Figure 4.3). The color of the metal precipitates in the effluent tank
changed from dark brownish in the previous periods to turquoise light brown. These
observations suggested that alternative metal precipitates other than metal sulfides were
formed as the sulfide concentration constantly fluctuated during this reactor operation
period (Figure 4.2).

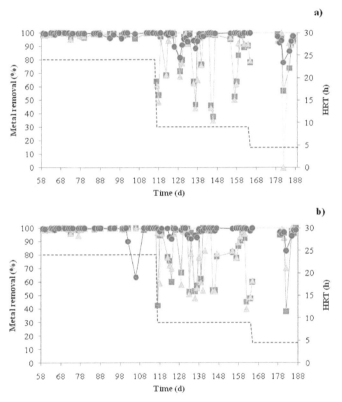

Figure 4.3 Cu (circles), Zn (squares) and Cd (triangles) removal efficiency in a) R1 and
b) R2 as a function of the HRT.

4.3.3 Speciation analysis

The thermodynamic model shows a wide diversity of oversaturated species in the
influent tank before entering the bioreactors, mainly as phosphate, hydroxide and sulfate
compounds (Table 4.2). However, the theoretical precipitates formed at equilibrium

were only hydroxyapatite ($Ca_5(PO_4)_3(OH)$) as well as crystalline (c) and amorphous (s) pyromorphite [$Pb_5(PO_4)_3Cl$] with a saturation index of 19 and 31.6-27.6, respectively (Table 4.2). These precipitates were included in the speciation model in the presence of sulfide, which predicted that only hydroxyapatite remained precipitated, while pyromorphite was transformed to galena (PbS) at sulfide concentrations exceeding the metal to sulfide stoichiometric ratio due to its very low solubility product (Table 4.2).

The sulfide concentration in the IFB reactor during the reactor operation at 24 and 9 h (Figure 4.1, Table 4.1) ensured the precipitation of the metals as covellite for Cu, Galena for Pb, greenockite for Cd and sphalerite and wurzite for Zn with a saturation index in descendant order and in agreement with their solubility product (Table 4.2). At low sulfide concentrations (<50 mg/L) such as the ones observed at an HRT of 4.5 h (Figure 4.2), several metal-sulfate and phosphate precipitates were likely to be produced instead of sulfide precipitates (Table 4.2). However, it should be noted that the model used is based on thermodynamics rather than kinetic data, and hence, some of the precipitates formed during the timescale of the experiment might not be present at the equilibrium. In accordance, XRD analysis of the precipitates formed at the three HRTs tested in the IFB reactor (data not shown) shows the presence of covellite and hydroxyapatite, while the other precipitates predicted in the speciation model were not observed probably due to their poor crystallinity.

4.3.4 Particle size distribution of the IFB reactor precipitates

The PSD of the metal precipitates during the three HRTs ranged from <0.5 to 250 μm, but the majority of the particles were between 1 to 50 μm (Figure 4.4). In R1, the sulfide concentration remained similar at the HRTs of 24 and 9 h, while the PSD showed important differences. The percentage of particles above 10 μm at the HRT of 24 h was substantially higher compared to the HRT of 9. Furthermore, the PSD shows a change in size range with the decrease of HRT from 24 to 9 h, where the amount of particles in the range of 50 to 100 and 100 to 250 μm decreased, while the number of particles below 0.5 μm increased (Figure 4.4). The PSD results were similar for both reactors despite the differences in sulfide concentration. The metal removal efficiency decreased at an HRT of 9 h for Zn and Cd (Figure 4.3), which was associated with the appearance of particles below 0.5 μm in both reactors: up to 10.3% for R1 and 9.4% for R2 (Figure 4.4). The sulfide concentration fluctuated strongly and decreased at the HRT of 4.5 h (Figure 4.4). At this HRT, the PSD displayed a wider range in size compared with the previous HRTs tested, but the range containing the highest percentage of particles was from 1-10 μm in both reactors.

Table 4.2 Oversaturated species predicted in the speciation model in the influent and the bioreactor liquor at the sulfide concentrations obtained at the different operational conditions. Indexed small table shows the solubility product (log K_{SP}) of the metal sulfide species formed at equilibrium. Number in parentesis indicate the saturation index and minerals highlighted in bold indicate the precipitates predicted at equilibrium.

Influent	Sulfide concentration (mg/L)			
(No sulfide)	34.9	50	290	441*
Chloropyromorphite$_{(c)}$ (**31.6**)	Antlerite (4.1)	Antlerite (4.2)	**Covellite (19.8)**	**Covellite (19.5)**
Chloropyromorphite$_{(s)}$ (27.6)	Atacamite (3)	Atacamite (3.3)	Cu$_3$(PO$_4$)$_{2(s)}$ (0.2)	**Galena (5.9)**
Hydroxylpyromorphite (19.0)	Brochantite (7.4)	Brochantite (7.6)	**Galena (6.5)**	**Greenockite (5.1)**
Pb$_3$(PO$_4$)$_{2(s)}$ (11.3)	Cd$_3$(PO$_4$)2$_{(s)}$ (1)	**Covellite (18.4)**	**Greenockite (5.8)**	**Hydroxyapatite (2)**
Tsumebite (7.6)	**Covellite (17.9)**	Cu(OH)$_{2(s)}$ (0.6)	**Hydroxyapatite (2)**	**Spharelite (6.7)**
Brochantite (5.8)	Cu(OH)$_{2(s)}$ (0.4)	Cu$_3$(PO$_4$)$_{2(s)}$ (6.2)	**Spharelite (7)**	Wurtzite (4.5)
Zn$_3$(PO$_4$)$_2$•4H$_2$O$_{(s)}$ (5.6)	Cu$_3$(PO$_4$)$_{2(s)}$ (6)	Cu$_3$(PO$_4$)$_2$•3H$_2$O$_{(s)}$ (4.5)	Wurtzite (4.8)	
Cu$_3$(PO$_4$)$_{2(s)}$ (4.7)	Cu$_3$(PO$_4$)$_2$•3H$_2$O$_{(s)}$ (4.3)	**Galena (4.2)**		
Langite (3.5)	**Galena (4.2)**	**Greenockite (9.2)**		
PbHPO$_{4(s)}$ (3)	**Greenockite (9.4)**	**Hydroxyapatite (1.5)**		
Cu$_3$(PO$_4$)$_2$•3H$_2$O$_{(s)}$ (3)	**Hydroxyapatite (1.3)**	Langite (5.4)		
Antlerite (3)	Langite (5.2)	**Spharelite (7.6)**		
Atacamite (2.1)	**Spharelite (7)**	Tenorite(am) (1.4)		
Cd$_3$(PO$_4$)2$_{(s)}$ (2.6)	Tenorite(am) (1.2)	Tenorite(c) (2.2)		
Larnakite (2.8)	Tenorite(c) (2.1)	**Wurtzite (5.4)**	Name	log K_{SP}
Tenorite$_{(c)}$ (1.6)	**Wurtzite (4.9)**	Zn$_3$(PO$_4$)$_2$•4H$_2$O$_{(s)}$ (6.6)	Covellite (CuS)	-22.22
Pb(OH)$_{2(s)}$ (1.1)	Zn$_3$(PO$_4$)$_2$•4H$_2$O$_{(s)}$ (6.5)		Galena(PbS)	-14.92
Pb$_3$O$_2$SO$_{4(s)}$ (0.9)			Greenockite (CdS)	-14.02
Tenorite$_{(am)}$ (0.8)			Sphalerite (ZnS)	-10.82
Hydroxyapatite (0.7)			Wurtzite (ZnS)	-8.62
Pb$_2$(OH)$_3$Cl$_{(s)}$ (0.7)				

* Precipitates and values did not change above this sulfide concentration, (c) crystal, (s) solid and (am) amorphous.

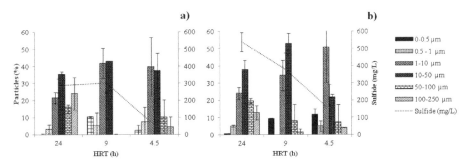

Figure 4.4 PSD and sulfide concentration for a) R1 and b) R2 at the HRTs evaluated.

4.3.5 Metal depletion kinetics in the batch experiments

Metal precipitation experiments were done to differentiate metal-sulfide formation-precipitation (filtrated samples) from metal-sulfide settling (samples without filtration). The metal depletion rate of the metals in the samples without filtration occurred in the following order: Cu and Pb depleted first, followed by Zn and Cd. After 8 h metals were still detected in the liquid phase (Figure 4.5a) suggesting that the metal sulfides formed were not big and thus, not heavy enough to settle. The metal concentrations in the filtrated samples depleted immediately (Figure 4.5b). Zn, Cd, Pb and Cu concentrations after the first measurement were 14.2, 16.1, 29.4 and 24.6 mg/L, respectively. However, fluctuations in metal concentrations ranging from 20-40 mg/L were observed until the end of the experiment (Figure 4.5b). These fluctuations were probably due to the size of the metal sulfides formed, which were still able to pass through the 0.45 μm filter used for metal analysis.

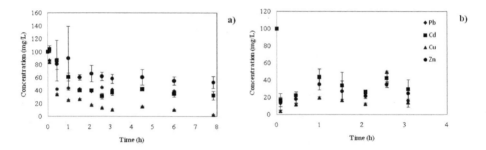

Figure 4.5 Metal concentration in the liquid phase from the batch experiments as a function of time on the samples (a) without and (b) with filtration prior to acidification.

4.3.6 SRA in the biofilm and in the bioreactor liquid

The specific SRA and biomass concentration in R1 was lower than in R2 in both the biofilm and bioreactor liquid samples (Figure 4.6). This could be attributed to the higher COD and sulfate concentrations supplied to R2 compared to R1. After metal addition, the specific SRA and biomass concentration in the biofilm decreased in both reactors. In R1, the specific SRA decreased from 44.9 to 21.9 mg S^{2-}/gVSS*h and in R2 from 95.2 to 69.1 mg S^{2-}/gVSS*h, while in the bioreactor liquid samples, the specific SRA remained similar for R1 before and after metal addition (23.4 mg S^{2-}/gVSS*h) and increased from 47.8 to 62.6 mg S^{2-}/gVSS*h in R2. The biomass concentration in the bioreactor liquid samples (0.2 gVSS/L) remained similar before and after metal addition.

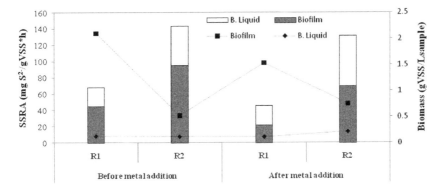

Figure 4.6 Specific sulfate reducing activity (□ □) and biomass concentration (■ ,♦) of the biofilm and bioreactor liquid prior and after metal addition.

At the end of the reactor operation, no activity in the biofilm and bioreactor liquid was observed and therefore, the results are not shown. The sulfide concentration in batch experiments with the bioreactor liquid and biofilm samples at the end of the reactors operation displayed only sulfide values for R1 (76.8 mg S^{2-}/L) after 333 h of incubation, yielding similar values as in the controls with no COD added.

4.3.7 Microbial community development
The biofilm samples in both reactors showed a similar increase of the quantitative composition of the SRB community and reached a similar maximum value close to 650 SRB gene copies/µL (Figure 4.7). After increasing the COD concentration in R2 on day 59, an increase in the SRB concentration together with a decrease in the total bacterial number was observed in the biofilm in both reactors. In general, after metals addition and prior to the change in HRT to 9 h, the biofilm displayed a larger SRB population than in the bioreactor liquid in both reactors (610 versus 1587 SRB gene copies/µL, Figure 4.7). However, the SRB community in the bioreactor liquid displays a different trend during the different operational conditions: the SRB gene copies in the biofilm decreased with metal addition and HRT change, whereas the SRB gene copies in the suspended biomass from the bioreactor liquid increased during the whole reactor

operation except after the decrease in HRT to 4.5 h (Figure 4.7). At the latter HRT, the less affected SRB community was the one from R1 in the bioreactor liquid (Figure 4.7). The total bacterial number followed the trends of the SRB community upon the operational changes applied to the reactor, except for the change in HRT to 9 h where the total bacterial number in the liquid phase highly increased in both reactors (376000 and 5.91000 total bacterial gene copies/ µL for R1 and R2, respectively).

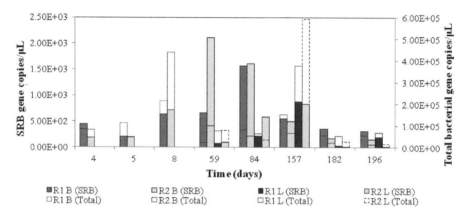

Figure 4.7 Quantification of the SRB community (primary axis) and total bacterial community (secondary axis) in the biofilm and bioreactor liquid at the days of the operational changes applied.

4.4 Discussion

4.4.1 Metal precipitation characteristics

This study shows that the HRT is an important factor determining the size of the metal sulfide precipitates (Figure 4.4) as this operational parameter affects the retention time of the metal sulfides for growth and the sulfide production for metal sulfide precipitation (Figure 4.2). The size of the precipitates (Figure 4.4) determines the settling rate of the metal sulfides (Figure 4.5), and consequently, the metal removal efficiencies (Figure 4.3). Other operational parameters that could affect the size of the metal sulfides due to changes in solubility, such as the pH (Villa-Gomez et al. 2012a; Mokone et al. 2010), remained fairly stable at neutral values in both reactors (Figure 4.1) and hence, their influence can be discarded. Metal sulfides are highly insoluble at neutral pH (Table 4.2), and only strong drops in pH to as low as 5, for ZnS, CdS and PbS, and as low as 2 for CuS can increment their solubility (Peters et al. 1984).

At HRTs of 24 and 9 h, the sulfide concentration in both reactors exceeded the metal to sulfide stoichiometric ratio (Figure 4.2, Table 4.1). This resulted in metals precipitating as sulfides (Table 4.2) and high supersaturation levels were maintained in both reactors favoring nucleation over crystal growth (Al-Tarazi et al. 2004; Mokone et al. 2010; Villa-Gomez et al. 2011b; Lewis 2010). Therefore, the variations in particle size were mainly due to the residence time of the precipitates allowing agglomeration of the small precipitates. An indication of agglomeration/aggregation is the low percentage of particles in the small size ranges, followed by a higher percentage of the mid-size ranges and then again a small percentage of bigger size ranges (Figure 4.4). The former are due to an increment of the number of large particles at the expense of the number of mid-size particles as a function of the residence time (Lewis 2010).

At an HRT of 9 h, metal precipitates at smaller ranges were observed (Figure 4.4) and at the same operational conditions, metal removal efficiencies decreased for Cd and Zn (Figure 4.3). This is an indication that the HRT was too short to allow all the Cd and Zn sulfide precipitates to grow and settle. Although the time scales of particle growth in metal sulfide precipitation has only been reported for a retention time of 200 min in chemical precipitator reactors (Frédéric et al. 2009), in contrast with a change of retention time from 24 to 9 h applied in this study (Figure 4.4), there is evidence that the growth of the precipitates continues exponential at the time length studied in chemical precipitator reactors (Frédéric et al. 2009; Couto and Mesquita 1994). In addition, substances commonly present in bioreactors such as phosphate (Villa-Gomez et al. 2012b) and bioproducts such as extracellular proteins (Moreau et al. 2007), which are not present in chemical precipitator reactors, could enhance the agglomeration process. The presence of extracellular proteins produced by SRB indeed promotes the aggregation of zinc sulfide precipitates in natural environments (Moreau et al. 2004; Moreau et al. 2007).

The change of HRT to 4.5 h showed that this operation mode is a risk for the process performance as the sulfide production highly decreased (Figure 4.2). Although lower sulfide concentrations could have led to a decrease in supersaturation, which is favorable for crystal growth of metal sulfides (Mersmann 1999), the formation of alternative precipitates such as brochantite ($Cu_4(OH)_6SO_4$) (Mokone et al. 2010) and Zn-phosphate (Villa-Gomez et al. 2012b) also occurred (Table 4.2). Therefore, the decrease of the HRT should be aligned to the sulfate reduction rate in order to maintain an adequate sulfide concentration within the reactors ensuring metal sulfide precipitation, yet avoiding free metal toxicity.

4.4.2 Settling characteristics of the metal sulfide precipitates

The settling rate of the metals obtained in the batch experiments (Figure 4.5a) followed the solubility product order (Table 4.2). Accordingly, CuS precipitates settle faster than other metals such as Fe (Ucar et al. 2011) and Zn (Sahinkaya et al. 2009), probably due to their higher tendency to agglomerate (Mokone et al. 2010; Sahinkaya et al. 2009; Villa-Gomez et al. 2012c). This is because nucleation particles tend to agglomerate more in comparison with crystals (Mersmann 1999), and the lower solubility product value (Table 4.2), the higher the number of these particles (Al-Tarazi et al. 2004; Lewis 2010).

Previous authors have used several settling tanks to separate metals based on their differences in settling rate (1992) as well as on the manipulation of other operational parameters such as the pH (Ucar et al. 2011). Due to its configuration, the use of the IFB reactor is a step forward to this approach as it acts not only as a sulfate reducing (sulfide producing) bioreactor but also as a settler for recovery of CuS precipitates, whereas the remaining metal sulfides with higher solubility product can be recovered in subsequent settlers.

4.4.3 Variations of the SRB activity and community

The effect of the operational changes was observed by tracking the variations of the SRB activity (Figure 4.6) and community (Figure 4.7) in the biofilm and the bioreactor liquid in the IFB bioreactors. The specific SRA (Figure 4.6) and the number of SRB copies (Figure 4.7) were higher in R2 than in R1 prior to the operational changes due to the higher COD and electron acceptor concentration (Figure 4.1) that favored bacterial (especially SRB) growth (Oyekola et al. 2009). In addition, the quantitative composition of the SRB community displayed a noticeable disparity between the biofilm and the biomass in the bioreactor liquid (Figure 4.7), which demonstrates that the biofilm was the main responsible for the sulfide production in both bioreactors.

Despite that the sulfide concentration produced in both IFB reactors (Table 4.2) prevented free metal toxicity (Kaksonen and Puhakka 2007), and that biofilms are reported to be more resistant towards high metal concentrations compared to suspended biomass (van Hullebusch et al. 2003), in this study, the biofilm was more affected by the metal addition (Figure 4.6) and by the HRT decrease (Figure 4.7) compared to the biomass from the bioreactor liquid. This could be due to the tendency of the metal sulfides to precipitate within/near where the sulfide production mainly occurs and therefore, local supersaturation is encountered (Villa-Gomez et al. 2011a), causing, consequently, inhibition of the SRB activity (Utgikar et al. 2002).

Despite the significant reduction of the specific SRA of the biofilm in both reactors following metal addition (Figure 4.6), no significant relationship between the specific SRA, the abundance of the SRB community (Figure 4.7) and reactor performance (Table 4.1, Figure 4.2) was observed. This suggests that the SRB population was only deactivated but still viable. Similarly, Uitkar et al. (Utgikar et al. 2002) observed that the SRB cultures were still viable after a strong decrease in their activity in a study about metal toxicity by insoluble metal sulfides in batch reactors. In contrast, the size of the SRB biofilm community itself decreased after the change in HRT to 9 (Figure 4.7). This decrease could be due to an excessive accumulation of the metal sulfides causing mortality. Notwithstanding this possibility, the presence of metal sulfide precipitates at sizes below 0.5 µm (Figure 4.4), especially for Zn and Cd (Figure 4.3), which were not observed at the HRT of 24 h, could have contributed to the decreased size of the SRB community. The presence of small ZnS precipitates, even at nanometer scale, has been previously postulated in bioreactors (Dar et al. 2009) and in natural environments (Moreau et al. 2007), however, the correlation between these metal sulfide sizes and their effects on SRB activity/community has not been studied yet. Therefore, further research using techniques to distinguish the size of the precipitates that are located in the biomass, such as scanning electron microscopy (Hennebel et al.) and transmission electron microscopy, should be used to confirm the correlation between the particle size of the metal sulfides and the SRB activity/population.

The change in the HRT to 4.5 h caused a significant performance drop (Table 4.1), indicating that the loading rate threshold may have passed the sulfate reduction rate, as indicated by the decrease in the SRB activity (Figure 4.6) and community (Figure 4.7). In agreement, by studying the kinetics of sulfate reduction and lactate utilization in chemostat cultures, Oluwaseun et al. (Oluwaseun et al. 2011) observed that sulfate reduction rates increased at starting sulfate concentrations of 2.5 and 5.0 g/L and long residence times (3-5 d) yielding sulfide concentrations of 0.13-0.53 g/L, whereas at short residence times (1-2 d), associated with sulfide concentrations of 0.014-0.088 g/L, there was a sharp decline in the volumetric sulfate reduction rates. In the aforementioned study, however, no metals were added and thus, the decrease of the sulfate reduction rate was exclusively due to the change in electron acceptor and donor loading rates. Bioreactor operation has been reported at HRTs as short as 6.5 h (Kaksonen et al. 2004) for less toxic metals (Fe and Zn) or at lower metal concentrations (Gallegos-Garcia et al. 2009). Therefore, this study is an important step forward to gain insight into the effect of bioreactor operation at low HRTs and high metal loading rates on the SRB community.

The lack of SRB activity (Figure 4.6) and the strong decrease of the SRB community (Figure 4.7) at the end of the reactor operation supports that free metal toxicity had occurred and demonstrated that the process failure was irreversible. Indeed, the metal effluent concentrations (Figure 4.3) were above the toxic values reported for SRB in both bioreactors (Kaksonen et al. 2006; Gonzalez-Silva et al. 2009). Zn completely inhibits SRB in the concentrations range 25-60 mg/L (Kaksonen et al. 2006), while 337.2 mg/L of Cd causes 44% inhibition of their substrate utilization rate in an UASB reactor (Gonzalez-Silva et al. 2009). It should be noted that the COD removal still proceeded after no SRB activity was observed anymore (Table 4.1) and a severe decrease in the SRB community (Figure 4.7). This can be attributed to the presence of other different microbial groups (Figure 4.7) in the bioreactors, such as acetogens and fermentatives (Omil et al. 1998; Gallegos-Garcia et al. 2009).

4.5 Conclusions

- The change in HRT from 24 to 9 h affects the size of the metal sulfide precipitates due to the change in contact time allowing aggregation and growth of metal sulfidic precipitates.

- The further HRT decrease to 4.5 h affects the sulfate reducing activity for sulfide production and hence, the supersaturation level and solid phase speciation. As a consequence, this induced free metal toxicity and process failure. Moreover, metal precipitates other than metal sulfides were formed.

- Metal sulfides affect the SRB activity and community in the biofilm, probably because of the stronger local supersaturation causing metal sulfides accumulation in the biofilm.

References

Al-Tarazi M, Heesink ABM, Azzam MOJ, Yahya SA, Versteeg GF (2004) Crystallization kinetics of ZnS precipitation; an experimental study using the mixed-suspension-mixed-product-removal (MSMPR) method. Crystal Research and Technology 39 (8):675-685. doi:10.1002/crat.200310238

Al-Tarazi M, Heesink ABM, Versteeg GF (2005a) Effects of reactor type and mass transfer on the morphology of CuS and ZnS crystals. Crystal Research and Technology 40 (8):735-740

Al-Tarazi M, Heesink ABM, Versteeg GF, Azzam MOJ, Azzam K (2005b) Precipitation of CuS and ZnS in a bubble column reactor. AIChE Journal 51 (1):235-246

APHA APHA (2005) Standard methods for examination of water and wastewater. 20 edn., Washington D.C.

Bialek K, Kim J, Lee C, Collins G, Mahony T, O'Flaherty V (2011) Quantitative and qualitative analyses of methanogenic community development in high-rate anaerobic bioreactors. Water Research 45 (3):1298-1308. doi:10.1016/j.watres.2010.10.010

Bijmans MFM, van Helvoort P-J, Dar SA, Dopson M, Lens PNL, Buisman CJN (2009) Selective recovery of nickel over iron from a nickel-iron solution using microbial sulfate reduction in a gas-lift bioreactor. Water Research 43 (3):853-861

Boonstra J, van Lier R, Janssen G, Dijkman H, Buisman CJN (1999) Biological treatment of acid mine drainage. In: Amils R, Ballester A (eds) Process Metallurgy, vol Volume 9. Elsevier, pp 559-567

Cord-Ruwisch R (1985) A quick method for the determination of dissolved and precipitated sulfides in cultures of sulfate-reducing bacteria. Journal of Microbiological Methods 4 (1):33-36

Couto MS, Mesquita ON (1994) Nucleation and aggregation of ZnS crystallites precipitated from a chemcial reaction. Brazilian journal of physics 24 (2)

Dar S, Bijmans M, Dinkla I, Geurkink B, Lens P, Dopson M (2009) Population Dynamics of a Single-Stage Sulfidogenic Bioreactor Treating Synthetic Zinc-Containing Waste Streams. Microbial Ecology 58 (3):529-537. doi:10.1007/s00248-009-9509-9

Esposito G, Veeken A, Weijma J, Lens PNL (2006) Use of biogenic sulfide for ZnS precipitation. Separation and Purification Technology 51 (1):31-39

Ferris FG, Schultze S, Witten TC, Fyfe WS, Beveridge TJ (1989) Metal Interactions with Microbial Biofilms in Acidic and Neutral pH Environments. Applied and Environmental Microbiology 55 (5):1249-1257

Foti M, Sorokin DY, Lomans B, Mussman M, Zacharova EE, Pimenov NV, Kuenen JG, Muyzer G (2007) Diversity, activity, and abundance of sulfate-reducing bacteria in saline and hypersaline soda lakes, vol 73. vol 7.

Frédéric G, Kamal M-BM, Cournil M (2009) Precipitation dynamics of zinc sulfide multiscale agglomerates. AIChE Journal 55 (10):2553-2562

Fu F, Wang Q (2011) Removal of heavy metal ions from wastewaters: A review. Journal of Environmental Management 92 (3):407-418. doi:10.1016/j.jenvman.2010.11.011

Gallegos-Garcia M, Celis LB, Rangel-Méndez R, Razo-Flores E (2009) Precipitation and recovery of metal sulfides from metal containing acidic wastewater in a sulfidogenic down-flow fluidized bed reactor. Biotechnology and Bioengineering 102 (1):91-99

Gonzalez-Silva BM, Briones-Gallardo R, Razo-Flores E, Celis LB (2009) Inhibition of sulfate reduction by iron, cadmium and sulfide in granular sludge. Journal of Hazardous Materials 172 (1):400-407. doi:10.1016/j.jhazmat.2009.07.022

Hennebel T, De Gusseme B, Boon N, Verstraete W (2009) Biogenic metals in advanced water treatment. Trends in Biotechnology 27 (2):90-98

Kaksonen AH, Franzmann PD, Puhakka JA (2004) Effects of hydraulic retention time and sulfide toxicity on ethanol and acetate oxidation in sulfate-reducing metal-precipitating fluidized-bed reactor. Biotechnology and Bioengineering 86 (3):332-343

Kaksonen AH, Plumb JJ, Robertson WJ, Riekkola-Vanhanen M, Franzmann PD, Puhakka JA (2006) The performance, kinetics and microbiology of sulfidogenic fluidized-bed treatment of acidic metal- and sulfate-containing wastewater. Hydrometallurgy 83 (1-4):204-213

Kaksonen AH, Puhakka JA (2007) Sulfate Reduction Based Bioprocesses for the Treatment of Acid Mine Drainage and the Recovery of Metals. Engineering in Life Sciences 7 (6):541-564

Kaksonen AH, Riekkola-Vanhanen ML, Puhakka JA (2003) Optimization of metal sulphide precipitation in fluidized-bed treatment of acidic wastewater. Water Research 37 (2):255-266

Lee C, Kim J, Hwang K, O'Flaherty V, Hwang S (2009) Quantitative analysis of methanogenic community dynamics in three anaerobic batch digesters treating different wastewaters. Water Research 43 (1):157-165. doi:10.1016/j.watres.2008.09.032

Lens PNL, Hulshoff Pol L, Wilderer P (2002) Water Recycling and Resource Recovery in Industry: Analysis, Technologies and Implementation. Integrated environmental Technology Series. IWA Publishing, Cornwall

Lewis AE (2010) Review of metal sulphide precipitation. Hydrometallurgy 104 (2):222-234

Magot M, Caumette P, Desperrier JM, Matheron R, Dauga C, Grimont F, Carreau L (1992) Desulfovibrio longus sp. nov., a sulfate-reducing bacterium isolated from an oil-producing well. International journal of systematic bacteriology 42 (3):398-403

Mersmann A (1999) Crystallization and precipitation. Chemical Engineering and Processing 38 (4-6):345-353

Mishra PK, Das RP (1992) Kinetics of zinc and cobalt sulphide precipitation and its application in hydrometallurgical separation. Hydrometallurgy 28 (3):373-379

Mokone TP, van Hille RP, Lewis AE (2010) Effect of solution chemistry on particle characteristics during metal sulfide precipitation. Journal of Colloid and Interface Science 351 (1):10-18. doi:10.1016/j.jcis.2010.06.027

Moreau JW, WEBB, I. R, BANFIELD, F. J (2004) Ultrastructure, aggregation-state, and crystal growth of biogenic nanocrystalline sphalerite and wurtzite, vol 89. vol 7. Mineralogical Society of America, Washington, DC, ETATS-UNIS

Moreau JW, Weber PK, Martin MC, Gilbert B, Hutcheon ID, Banfield JF (2007) Extracellular Proteins Limit the Dispersal of Biogenic Nanoparticles. Science 316 (5831):1600-1603. doi:10.1126/science.1141064

Neculita C-M, Zagury GJ, Bussière B (2008) Effectiveness of sulfate-reducing passive bioreactors for treating highly contaminated acid mine drainage: I. Effect of hydraulic retention time. Applied Geochemistry 23 (12):3442-3451

Oluwaseun O, Oyekola, Robert P, Hille v, Harrison STL Kinetic Description of the Competitive Interaction between Lactate Oxidizers and Fermenters in a Biosulfidogenic System In: World Congress on Engineering and Computer Science San Francisco, Ca., 2011. International Association of Engineers

Omil F, Lens P, Visser A, Hulshoff Pol LW, Lettinga G (1998) Long-term competition between sulfate reducing and methanogenic bacteria in UASB reactors treating volatile fatty acids. Biotechnology and Bioengineering 57 (6):676-685. doi:10.1002/(sici)1097-0290(19980320)57:6<676::aid-bit5>3.0.co;2-i

Oyekola OO, van Hille RP, Harrison STL (2009) Study of anaerobic lactate metabolism under biosulphidogenic conditions. Water Research 43: 3345-3354

Peters RW, Chang T-K, Ku Y (1984) Heavy metal crystallization kinetics in an MSMPR crystallizer employing sulfide precipitation. Journal Name: AIChE Symp Ser; (United States); Journal Volume: 80:240:Medium: X; Size: Pages: 55-75

Sahinkaya E, Gungor M, Bayrakdar A, Yucesoy Z, Uyanik S (2009) Separate recovery of copper and zinc from acid mine drainage using biogenic sulfide. Journal of Hazardous Materials 171 (1-3):901-906

Sampaio RMM, Timmers RA, Xu Y, Keesman KJ, Lens PNL (2009) Selective precipitation of Cu from Zn in a pS controlled continuously stirred tank reactor. Journal of Hazardous Materials 165 (1-3):256-265

Siggins A, Enright AM, O'Flaherty V (2011) Temperature dependent (37-15°C) anaerobic digestion of a trichloroethylene-contaminated wastewater. Bioresource Technology 102 (17):7645-7656

Tabak HH, Scharp R, Burckle J, Kawahara FK, Govind R (2003) Advances in biotreatment of acid mine drainage and biorecovery of metals: 1. Metal precipitation for recovery and recycle. Biodegradation 14 (6):423-436

Ucar D, Bekmezci OK, Kaksonen AH, Sahinkaya E (2011) Sequential precipitation of Cu and Fe using a three-stage sulfidogenic fluidized-bed reactor system. Minerals Engineering 24 (11):1100-1105. doi:10.1016/j.mineng.2011.02.005

Utgikar VP, Harmon SM, Chaudhary N, Tabak HH, Govind R, Haines JR (2002) Inhibition of sulfate-reducing bacteria by metal sulfide formation in bioremediation of acid mine drainage. Environmental Toxicology 17 (1):40-48

van Hullebusch E, Zandvoort M, Lens P (2003) Metal immobilisation by biofilms: Mechanisms and analytical tools. Reviews in Environmental Science and Biotechnology 2 (1):9-33

Veeken AHM, de Vries S, van der Mark A, Rulkens WH (2003) Selective Precipitation of Heavy Metals as Controlled by a Sulfide-Selective Electrode. Separation Science and Technology 38 (1):1 - 19

Villa-Gomez D, Ababneh H, Papirio S, Rousseau DPL, Lens PNL (2011a) Effect of sulfide concentration on the location of the metal precipitates in inversed fluidized bed reactors. Journal of Hazardous Materials 192 (1):200-207. doi:10.1016/j.jhazmat.2011.05.002

Villa-Gomez D, Papirio S, Van Hullebusch E, Farges F, Nikitenko S, Kramer H, Lens PNL (2011b) Influence of sulfide concentration on metal precipitate characteristics for potential metal recovery. Submitted to Bioresource Technology

Villa-Gomez DK, Maestro R, van Hullebusch ED, Farges F, Nikitenko S, Kramer H, Gonzalez-Gil G, Lens PNL (2012a) Effect of pH on the characteristics of the metal sulfide precipitates from sulfide produced in bioreactors. Submitted to Environmental Science and Technology

Villa-Gomez DK, Papirio S, van Hullebusch ED, Farges F, Nikitenko S, Kramer H, Lens PNL (2012b) Influence of sulfide concentration and macronutrients on the characteristics of metal precipitates relevant to metal recovery in bioreactors. Bioresource Technology (0). doi:10.1016/j.biortech.2012.01.041

Villa-Gomez DK, Papirio S, van Hullebusch ED, Farges F, Nikitenko S, Kramer H, Lens PNL (2012c) Influence of sulfide concentration and macronutrients on the characteristics of metal precipitates relevant to metal recovery in bioreactors. Bioresource Technology 110 (0):26-34. doi:10.1016/j.biortech.2012.01.041

Wagner M, Loy A, Klein M, Lee N, Ramsing NB, Stahl DA, Friedrich MW (2005) Functional Marker Genes for Identification of Sulfate-Reducing Prokaryotes. In: Jared RL (ed) Methods in Enzymology, vol Volume 397. Academic Press, pp 469-489. doi:http://dx.doi.org/10.1016/S0076-6879(05)97029-8

Yu Y, Lee C, Kim J, Hwang S (2005) Group-specific primer and probe sets to detect methanogenic communities using quantitative real-time polymerase chain reaction. Biotechnology and Bioengineering 89 (6):670-679. doi:10.1002/bit.20347

Zehnder AJB, Huser BA, Brock TD, Wuhrmann K (1980) Characterization of an acetate-decarboxylating, non-hydrogen-oxidizing methane bacterium. Archives of Microbiology 124 (1):1-11. doi:10.1007/bf00407022

CHAPTER

5

TUNING STRATEGIES

to control

the sulfide concentration

USING A pS ELECTRODE

In the IFB

REACTOR

Abstract

Tuning strategies to control the dissolved sulfide concentration in a sulfate reducing bioreactor using a direct measurement of the sulfide species (pS electrode) were evaluated. The experiments were performed in an inverse fluidized bed (IFB) bioreactor with automated operation using LabVIEW software version 2009®. Step changes in the OLR through variations in the influent chemical oxygen demand (COD_{in}-Tuning I) concentration or in the hydraulic retention time (HRT-Tuning II) at constant COD/SO_4^{2-} ratio (0.67) were applied to create a sulfide response. The pS output values obtained from tuning I and II were used to determine proportional-integral-derivative (PID) controller parameters. A rapid response and high sulfide increment was obtained through an increase in the COD_{in} concentration, while a decrease in the HRT exhibited a slower response and a smaller sulfide increment. Decreasing the OLR irrespective of the tuning strategy caused a longer response time of the bioreactor and a poor or non observable decrease of the sulfide concentration due to sulfide accumulation (tuning II) or non-accounted substrate sources utilization (tuning I). The pS electrode response is adequate for its application in controlling sulfate reducing bioreactors. However, pH variations and high sulfide concentrations should be carefully observed for correction of the pS values displayed.

This Chapter has been accepted for publication as:

Villa-Gomez D. K., Cassidy J., Keesman K., Sampaio R. and Lens P.N.L. (2013) Strategies for sulfide control using a pS electrode in sulfate reducing bioreactors. Water Research. In press.

5.1 Introduction

Biological sulfate reduction is a process for the treatment of metal containing wastewaters enabling the recovery of metals as sulfidic precipitates (Bijmans et al. 2011). Sulfate reducing bacteria (SRB) reduce sulfate through the oxidation of either organic compounds or hydrogen, resulting in the production of hydrogen sulfide (Kaksonen and Puhakka 2007). Most of the metal containing wastewaters are deficient in organic compounds (Papirio et al. 2012). Thus, their addition as electron donor for sulfate reduction determines the overall costs of the process (Gibert et al. 2004; Zagury et al. 2006). For metal removal and recovery processes, the required amount of sulfide to be produced by SRB depends on the composition of the wastewater to be treated, i.e. its metal concentration. Steering the sulfide production towards this required stoichiometric amount in bioreactors is highly relevant to avoid over production of H_2S that increases operational costs and may impose a sulfide removal post-treatment step.

Process control has been used for several biological production processes yielding desirable end products such as ethanol, penicillin and diverse fermentation products as well as for waste water treatment (Dunn et al. 2005). In these processes, typically, set-point control is based on the manipulation of temperature, pH, substrate or dissolved oxygen concentration. In anaerobic digestion, control variables commonly used for process control are intermediate compounds such as volatile organic acid concentration, pH, bicarbonate alkalinity or gas concentrations/flow rates (Pind et al. 2003).

Even though large progress has been made in the control of anaerobic (methanogenic) systems, there is insufficient knowledge about process control of sulfate reducing bioreactors. Mathematical models have been developed to support the design of a control strategy for sulfate reduction in bioreactors (Kalyuzhnyi and Fedorovich 1998; Oyekola et al. 2012; Gupta et al. 1994). In these studies the objective was to outcompete or favor microbial trophic groups other than SRB, while accounting for the control of the sulfide production. Torner-Morales and Buitrón (2010) used the redox potential as a control variable to maintain the sulfate reduction efficiency and subsequent partial sulfide oxidation in a single sequencing batch reactor unit. This controlled variable, combined with the control of the pH, allowed a combined sulfate-reducing/sulfide-oxidizing process with a continuous operation and a significant yield of elemental sulfur (64%).

The development of a control strategy based on the sulfide concentration as the controlled variable is a more direct approach than the redox potential for the control of the sulfide concentration for metal recovery. For this, the most adequate sensor developed is the pS electrode that measures the activity of the S^{2-} species. The use of a pS electrode in biological systems was already described by Dan et al. (1985) to monitor photosynthetic sulfide oxidation by *Chlorobium phaeobacteroides*. This sensor, in combination with a pH electrode, has also been successfully validated in the control of the sulfide concentration in precipitator reactors for selective metal recovery using chemically and biologically produced (biogenic) sulfide (König et al. 2006; Veeken et al. 2003b; Grootscholten et al. 2008; Veeken et al. 2003a; Sampaio et al. 2009).

The selection of an appropriate control strategy largely depends on the process characteristics. A proportional-integral (PI) control strategy using the pS electrode was found to be sufficient for biogenic sulfide control entering a precipitator reactor with only metal sulfide precipitation taking place (König et al. 2006). The control of the sulfide concentration directly in the SRB bioreactor is more complex as it needs to take into account the biological sulfide production process as well. Therefore, it requires a control strategy that manipulates the organic loading rate (OLR) to the sulfide producing SRB bioreactor, where the set-point ideally depends on the metal concentration that is required to be precipitated. Thus, an additional control parameter is required to overcome the lag time between substrate dosing and substrate degradation or transformation into the desired product. Even though robust control strategies appear as a very promising option over conventional control types such as the PI, there are only a few experimental results in the literature validating their application in anaerobic bioreactors (Steyer et al. 1999). The proportional-integral-derivative (PID) controller, which has been widely used in anaerobic bioreactors (Dunn et al. 2003; Dunn et al. 2005; Jagadeesh and Sudhaker 2010; Pind et al. 2003; Marsili-Libelli and Beni 1996), contains a derivative control parameter that is used to overcome lag phases, therefore, it can be seen as an option for sulfide control in SRB bioreactors. The PID control parameters can be obtained by using different tuning strategies and tested experimentally or through model simulations (Pind et al. 2003).

The aim of this study was thus to evaluate tuning strategies to manipulate the OLR via changing the influent COD (COD_{in}) concentration or the hydraulic retention time (HRT) to control the sulfide concentration in an inverse fluidized bed (IFB) reactor using a pS electrode and a PID controller. The evaluation of the tuning strategies was based on analyzing the response of the system to the applied change in terms of response time and time delay, load and set point (sulfide) changes, as well as robustness and stability of the sensor (Rodrigo et al. 1999). The Cohen-Coon method (Dunn et al. 2005) was used to determine the PID parameters of the controller. Since pS sensors have not yet been applied in sulfate reducing bioreactors, their feasibility for application was also analyzed.

5.2 Materials and methods

5.2.1 Reactor set-up

The experiments were carried out in an IFB bioreactor as described by Villa-Gomez et al. (2012) but with automated operation using a data acquisition card (NI cDAQ-9174, National Instruments, The Netherlands) and Labview software version 2009® (Figure 5.1). Lactate was used as electron donor and carbon source and sulfate was added as Na_2SO_4 at a COD/SO_4^{2-} ratio of 0.67. The synthetic medium used in the bioreactor experiment and batch activity tests was the same as used by Villa-Gomez et al. (2012).

The pH and pS in the IFB bioreactor were monitored using a sulfide resistant pH electrode (Prosense, Oosterhout, The Netherlands) and a solid state Ag_2S ion selective (S^{2-}) electrode (Prosense, Oosterhout, The Netherlands) of 40 cm length each inserted on the top of the bioreactor column (Figure 5.1). The Labview software® contained a PID controller (PID and Fuzzy Logic Toolkit, National Instruments, The Netherlands) for the control of the pH using stock solutions of HCl/NaOH connected to the recirculation tube and with PID parameters obtained by error minimization.

5.2.2 Experimental design

The IFB bioreactor was operated for 105 days at an HRT of 24 h with incremental changes to the COD_{in} concentration from 0.5 to 1.5 gCOD/L followed by a change in the HRT from 24 to 12 h on day 106 and, on day 133, the decrease in the COD_{in} concentration from 1.5 to 0.5 gCOD/L.

Then, several OLR step changes were applied in the IFB bioreactor to create responses of the sulfide concentration produced by SRB for the acquisition of the tuning coefficients of the PID controller. The experiments consisted of running the IFB bioreactor at a constant OLR and, once the pS electrode displayed constant values for at least 5 HRTs, to change this OLR to create a step response in the sulfide concentration until the time that the reactor reaches a new steady pS value (response time). The OLR was changed from 0.5 to 1 gCOD/L*d through either variations in the COD_{in} concentration (Tuning I) or in the HRT (Tuning II) by changing the influent flow rate. Once the reactor reached a new steady state in the pS values, the same methodology was applied but to return the OLR to its initial setting.

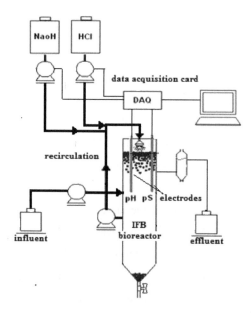

Figure 5.1. Experimental set-up of the IFB bioreactor with pH and pS online measurement as well as pH control.

The reliability of the pS electrode response under the IFB bioreactor operational conditions was also evaluated. The pS electrode response in the IFB bioreactor after sulfide variations ranging from 100 to 500 mg/L of total dissolved sulfide (TDS) concentration was analyzed. In addition, the pS signal was analyzed at TDS concentrations ranging from 20 mg/L to 400 mg/L at constant pH of 7 with chemically produced sulfide (Na_2S) and with biogenic sulfide. Chemically produced sulfide stock solutions were made with Na_2S*H_2O (Merck, extra pure, about 35% Na_2S), while biogenic sulfide samples were taken from the IFB bioreactor at different operation periods and thus, different TDS concentrations. In the chemically produced samples, TDS concentrations were re-measured once the pH was adjusted to 7, as the sulfide concentration modifies the pH. In the TDS samples from the bioreactor liquid, the pH was not adjusted as the bioreactor was already at pH 7 ($\pm$ 0.2). The theoretical S^{2-} was calculated using equation 1; these values were converted to mV with the calibration line made following the pS electrode calibration procedure.

$$S^{2-} = \frac{TDS}{1+\frac{(H^+)}{K_{a2}}+\frac{(H^+)^2}{K_{a2}K_{a1}}} \qquad (1)$$

Where: S^{2-} is measured by the pS electrode; $K_{a2} = 10^{-7}$ and $K_{a2} = 10^{-13.9}$

The pS electrode measures the concentration of the S^{2-} species (pS= -log $[S^{2-}]$), which depends on the TDS and pH (equation 1). Hence, an increase in the TDS leads to a decrease of the pS values. For practicality, sulfide in this manuscript refers to all sulfide species (HS, S^{2-} and H_2S), while TDS (all dissolved sulfide species) and pS (S^{2-}) are specific determinations of the sulfide species.

5.2.3 pS electrode calibration

The calibration of the pS electrode followed the procedure described by Veeken et al. (2003b), but using a sulfide solution containing the synthetic medium used in the bioreactor operation. Approximately 10 mM (320 mg/L) of Na_2S were titrated with 1 M of HCl, from high (12) to low (2.5) pH values. Two pS electrodes were used during this study: the first electrode exhibited a calibration curve that covered a range in voltage from 119 mV to 470 mV, while the working range of the second electrode covered from 353 to 672 mV. Both electrodes did not show significant variations in the repetitions of the calibration curve during this work and displayed a correlation coefficient > 0.99 (data not shown).

5.2.4 PID controller tuning

The pS electrode output values obtained from the step responses via tuning I and II were used to determine the tuning coefficients of the PID control. From these step responses the following tuning parameters were determined:

$$K = \frac{output\ (at\ steady\ state)}{input\ (at\ steady\ state)} = \frac{B}{A}$$

$$\tau = \frac{B}{S} \tag{2}$$

t_d = *time elapsed until the system responded (time dealy)*

Where B is equal to [pS_{final} - $pS_{initial}$], A is equal to [OLR_{final} - $OLR_{initial}$] and S is the slope of the sigmoidal response at the point of inflection.

The Cohen-Coon tuning method (Dunn et al. 2005) was used to provide estimates of the PID parameters (K_c, τ_i, τ_d) on the basis of the characteristics (equation 2) of the experimental step response data:

$$K_c = \frac{1}{K}\frac{\tau}{t_d}\left(\frac{4}{3} + \frac{t_d}{4\tau}\right) \qquad \tau_i = t_d\frac{32 + 6t_d/\tau}{13 + 8t_d/\tau} \qquad \tau_D = t_d\frac{4}{11 + 2t_d/\tau} \tag{3}$$

5.2.5 Residence time distribution in the IFB bioreactor

To discard time delays in the sulfide response due to the hydrodynamic behavior of the system, the residence time distribution (RTD) of the IFB bioreactor was determined as described by Warfvinge (2009) at a theoretical HRT of 12 and 24 h. The residence time distribution curves were determined using LiCl, since this compound is not degraded nor adsorbed by microorganisms (Olivet et al. 2005). LiCl was dissolved in small amounts of water and then injected at the top of the bioreactor column with a syringe over a time as short as possible. The amount of tracer used corresponded to a bulk concentration of 30 mg/L in the bioreactor working volume (2.5 L). Samples of the effluent were taken at predetermined time intervals until the recovery of the total tracer was completed.

5.2.6 Analyses

COD was measured by the closed reflux method (APHA 2005). Sulfate was measured as described by Villa-Gomez et al. (2011). TDS was determined spectrophotometrically by the colorimetric method described by Cord-Ruwish (1985) using a spectrophotometer (Perkin Elmer Lambda 20). Acetate was measured by gas chromatography (GC-CP 9001 Chrompack) after acidification of the samples with 5% concentrated formic acid and filtration through a 0,45 µm nitrocellulose filter (Millipore). The gas chromatograph was fitted with a WCOT fused silica column, the injection and detector temperatures were 175 and 300 °C, respectively. Methane was measured with a gas chromatograph (Gas Chromatograph CP 3800), fitted with a PORABOND column Q (25m*0,53mm*10μm) and a TCD detector. The carrier gas was helium at 15 psi, the oven temperature was 22°C, and the injection volume was 500 μL.

5.3 Results

5.3.1 Sulfide response during start-up of the IFB reactor

The TDS concentration increased after 9 days of the first change from 0.5 to 1 gCOD/L*d on day 14. The response time of the TDS concentration to the increase in OLR from 1 to 1.5 gCOD/L*d on day 47 took 2 days, reaching a maximum TDS value of 618 mg/L. After this maximum value, a strong drop of the TDS concentration to 222 mg/L was observed and later the TDS concentration began to increase again. Biomass lost from the system on day 89 due to clogging of one of the tubes caused an overflow in the reactor column and strongly decreased the TDS concentration (from 400 to 250 mg/L). The change of the HRT from 24 to 12 h on day 105 did not show noticeable changes in the TDS values, while the decrease in the COD_{in} concentration from 1.5 to 0.5 gCOD/L on day 133 caused the decrease of the TDS to values around 124 mg/L after 13 days of reactor operation.

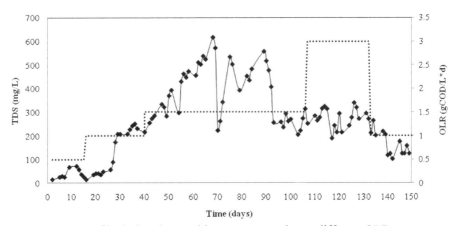

Figure 5.2. TDS profile during the IFB bioreactor operation at different OLR.

5.3.2 Evolution of the biomass activity in the IFB reactor

On day 40, the SSRA with lactate and acetate as electron donor was 32.5 and 1.7 gCOD/gVSS*d, respectively, while the SMA was 22.2 gCOD/gSSV*d (Table 5.1). After 150 days of the reactor operation (Figure 5.2), the SSRAs with lactate strongly decreased (12.5 gCOD/gVSS*d) probably due to the biomass lost on day 89, while the SSRA with acetate as electron donor highly increased (16.4 gCOD/gVSS*d). The SMA from day 146 onwards was no longer detectable (Table 5.1). The SSRA after 289 days of reactor operation notably increased for acetate (37.1 gCOD/gVSS*d) while for lactate, the increase was not that significant (Table 5.1).

Table 5.1. Evolution of the SSRA and SMA in the IFB reactor.

Time of the reactor operation (d)	SSRA lactate	SSRA acetate	SMA
		(gCOD/gVSS*d)	
40	32.5	1.7	22.23
146	12.5	16.4	ND
289	16.7	37.1	ND

ND: Not detectable

5.3.3 Response of the system to changes in OLR (tuning I and II)

In tuning I, the step changes in the OLR from 0.5 to 1 and from 1 to 0.5 gCOD/L*d increased and decreased, respectively the sulfate removal efficiency (Table 5.2), while the COD removal efficiency maintained similar values despite the OLR change (>80%). Prior to the first step response, no acetate could be detected, while the change in OLR to 1 and later to 0.5 gCOD/L*d caused an accumulation of acetate in the system to 28.2 and 52.2 mg/L, respectively. The TDS concentration increased from 113.4 to 333.1 mg/L with the change in OLR from 0.5 to 1 gCOD/L*d. The return to the initial OLR conditions yielded a different TDS concentration value at the steady state (152.8 mg/L) as compared to the value obtained on the first step change (Table 5.2).

Table 5.2 Steady state values in the IFB bioreactor prior to and after the change in OLR for tuning I and II (± deviation error, negligible for pS and pH).

	Tuning I			Tuning II		
	Change in COD$_{in}$			Change in HRT		
OLR (gCOD/L*d)	0.5	1	0.5	0.5	1	0.5
COD (g/L)	0.5	1	0.5	0.5	0.5	0.5
HRT (h)	24	24	24	24	12	24
Mean residence time (h)	21.7	21.7	21.7	21.7	11.15	21.7
SO$_4^{-2}$ removal (%)	39.3 ± 6.3	56.4 ± 13.9	24.6 ± 6.5	39.0 ± 5.4	63.2 ± 3.5	76.9 ± 15.1
COD removal (%)	86.3 ± 6.1	78.6 ±14	83.4 ± 6.5	63.8 ± 13.9	59.3 ± 5.5	75.9 ± 0.4
Acetate (mg/L)	0	28.2 ± 7.7	52.2	37.5 ± 6.5	121.6 ±20.1	34.2 ± 13.2
TDS (mg/L)	113.4 ± 15.68	333.1 ± 58.1	152.8± 6.5	131.0 ± 6.31	160.8 ± 12.8	207 ± 21.4
pH	6.8	7	6.9	6.9	6.7	6.8
pS	8.76	7.6	8.05	11.07	9.97	9.32
pS (mV)	-567	-592.7	-582.5	-513.7	-538.8	-553.8

The step response from 0.5 to 1 gCOD/L*d for tuning II caused an increase in both the TDS concentration (131 to 160.8 mg/L) and the sulfate removal efficiency (39 to 63.2%), while the return to the initial OLR value (0.5 gCOD/L*d) further increased both values (5. 2). The acetate concentration increased from 37.5 to 121.6 mg/L after the change in the OLR, while upon return to the initial OLR, acetate values were similar to the ones obtained prior to the step responses. The COD removal efficiency did not show noticeable changes after the changes in the OLR and only a slight increase (75.9%) was observed once the OLR returned to its initial value. The mean residence time in the IFB bioreactor for the theoretical HRT of 24 and 12 h was 21.7 and 11.15 h, respectively (Table 5.2). Figure 5.3 shows the RTD analysis of the IFB bioreactor that resembled the hydrodynamic behavior of a completely mixed reactor characterized by a rapid increase in the Li concentration in time followed by a slow, steady tail indicates a pulse moving through a completely mixed reactor (Figure 5.3). This behavior was confirmed for the IFB bioreactor operation at an HRT of 12 and 24 h.

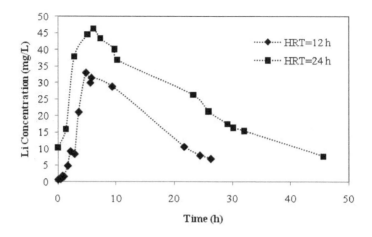

Figure 5.3 Residence time distribution profile in the IFB reactor.

5.3.4 pS electrode response to changes in OLR (tuning I and II)

The step response for tuning I displayed an exponential increase in the sulfide concentration that lasted 4 days with no significant time delay and reached a steady state pS value of 7.6 after the OLR increment (Figure 5.4a). When the OLR was again decreased to 0.5 gCOD/L*d, the pS response time was longer and the pS values at steady state (pS 8.0) were different compared to ones at the beginning of the experiment. Time delay in the response of the system was less than 24 h and the decrease of the sulfide concentration for this tuning strategy lasted 2 days until the values reached a steady state.

Tuning II showed a time delay of approximately 7 days and an exponential curve that lasted 15 days following the OLR step change (Figure 5.4b). The sulfide concentration slowly increased and consequently, the pS decreased from 11.1 to 9.9. The return of the HRT from 12 to 24 h showed a faster response in the system with a time delay of less than one day, reaching a steady state after 2 days. Surprisingly, the return to the initial conditions led to a further increase in the sulfide concentration giving a new pS value of 9.3 at steady state. During this experiment, strong drops in the pS values (sharp picks) were observed on day 7 and 30 due to fouling of the ion selective membrane that reduced its sensitivity. However, the electrode sensitivity was recovered immediately as this fouling was not-adhesive and thus, the bioreactor liquor contact cleans the membrane.

The pS signal showed in general negligible fluctuations in the signal for tuning I, while for tuning II, the signal fluctuated prior to the step change in OLR. This was due to small variations in the pH (Figure 5.3b), which induced changes in the S^{2-} species (Equation 1).

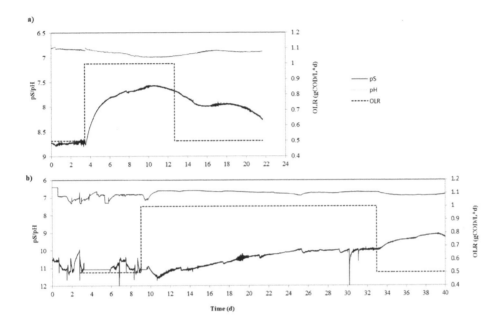

Figure 5.4 Step responses of the pS electrode response for a) tuning I and b) tuning II.

5.3.5 Tuning coefficients for the acquisition of the PID parameters

The pS response following the tuning I and II strategies (Figure 5.4) were analyzed for the acquisition of the tuning coefficients (Equation 2-Table 5.3) and PID parameters (Equation 3-Table 5.3). For comparison purposes, the response time and the ΔTDS are also included in Table 5.3. Tuning I showed the smallest response time with an important increment in the sulfide concentration for a step change in the OLR from 0.5 to 1 gCOD/L*d (Table 5.2), while in tuning II, a longer response time and a lower sulfide increment was observed for the same step change (Table 5.3).

The comparison of the coefficients obtained for tuning I and II show that the sulfide load change (K) for an increment in the OLR is higher in contrast to the values obtained when the OLR is decreased for both tuning strategies (Table 5.3). Furthermore, the value obtained in tuning II for a decrease in the OLR was positive, in contrast with the other K parameters. This was due to the increase in the sulfide concentration, despite the decrease in the OLR that lowers the substrate available for sulfate reduction. τ values display differences that are due to the differences in slope (Equation 1) of the exponential curve (Figure 5.4). The longest time delay was observed in tuning II for an increase in the OLR, while the shortest time delay was obtained in tuning I for a decrease in OLR (Table 5.3).

The comparison of the PID parameters showed that because of the negligible time delay (t_d), especially for the OLR decrease in tuning I, the contribution of the differential and integral part to the PID controller parameters was small, while the proportional gain (K_c) parameter was remarkably high, especially for the decrease in OLR as compared to the OLR increase in both tuning strategies (Table 5.3).

Table 5.3 Response time, TDS increment (ΔTDS), tuning coefficients and PID parameters obtained with tuning I and II.

	Tuning I		Tuning II (A)	
OLR step change (gCOD/L*d)	0.5-1	1-0.5	0.5-1	1-0.5
Response time (d)	4	2	15	2
ΔTDS	206.46	-180.3	29.8	46.17
K (L*d/g)	-2.23	-0.68	-2.18	1.27
τ (d)	30.83	-51.79	132.40	879.88
t_d (d)	0.29	0.04	2.00	0.40
K_c	-63.71	2420.57	-40.60	2311.62
τ_i	0.71	0.10	4.89	0.98
τ_D	0.10	0.01	0.73	0.14

5.3.6 Analysis of the pS electrode signal

Figure 5.5a shows a typical pS electrode response to a change in TDS concentration in the IFB bioreactor operation in the TDS concentration range between 100 and 50 mg/L. The pS response was consistent with the variations in TDS concentration and the small jumps in the signal on days 3.5, 6.5, 13 and 14.9 were due to small variations in pH (data not shown). Variations in the TDS concentration above 200 mg/L did not induce noticeable changes in the output pS value compared to the TDS concentrations below 200 mg/L. A similar trend was found in the pS response at low and high sulfide concentrations when the pS signal was tested to check its reliability at constant pH for biogenic and chemically produced sulfide (Figure 5.5b and 5.5c). The pS values show a logarithmic response with an exponential increase from approximately 0 to 100 mg/L of TDS (Figure 5.5b). Above this concentration, the pS values differed considerably between the biogenic and chemically produced sulfide as well as for the theoretical sulfide concentration (Figure 5.5c). This is because the differences in sulfide sampling between the sulfide sources (inside the bioreactor and in open bottles) are more noticeable at higher sulfide concentrations as the voltage response obtained from the S^{2-} concentration is less pronounced due to its logarithmic behavior (Equation 1) and thus, small variations in sulfide concentration highly affect the pS values.

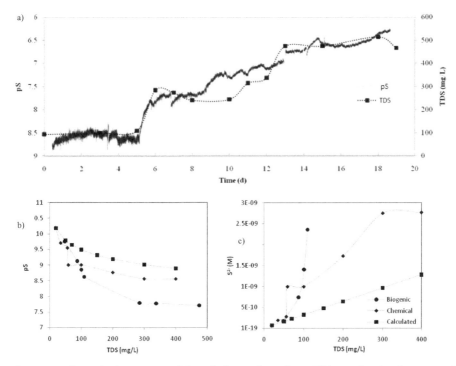

Figure 5.5 a) Typical response of the pS electrode to the sulfide production increment in the IFB bioreactor, b) Relationship between pS and TDS concentration c) Relationship between S^{2-} (M) and TDS concentration.

5.4 Discussion

5.4.1 Bioreactor response to tuning strategies applied

This study shows that the sulfide concentration in bioreactors can be controlled by the variation of the COD_{in} concentration (high ΔTDS) or the HRT (low ΔTDS) (Table 5.2) using the direct measurement of the sulfide species (S^{2-}) with a pS electrode (Figure 5.4). The PID parameters from the different step response experiments with both tuning strategies (Table 5.2) provide a baseline to evaluate the potential of the PID controller for its application in SRB bioreactors and to foresee, the parameters that need to be adapted or included (Heinzle et al. 1993).

At the beginning of the IFB reactor operation (Figure 5.2), time delays in the system after a change in the OLR occurred due to substrate utilization by microbial groups other than SRB such as methanogenic bacteria (Table 5.1) and lactate fermentation to acetate (Oyekola et al. 2012). However, the favorable thermodynamic (Widdel 1988) and operational conditions (Oyekola et al. 2009) for sulfate reduction that outcompeted the methanogenic activity (Table 5.1) decreased the response time at the latter stages of the IFB reactor operation. In addition, the increase in acetate utilization by SRB (Table 5.1) could contribute to the faster response time (Figure 5.2).

On the other hand, the differences in COD and sulfate removal efficiencies, sulfide and acetate production between tuning I and II for a positive or negative step change in OLR (Table 5.2) suggest that different metabolic pathways accounting for sulfate reduction and organic matter oxidation were induced (Dunn et al. 2005) depending on the tuning strategy applied. In addition, several days (response time) were needed to achieve pS steady values in both tuning strategies (Figure 5.4 and Table 5.3) that can destabilize the bioreactor due to an excesive control action needed. The substrate utilization by microbial groups other than SRB such as lactate fermentation to acetate (Oyekola et al. 2012) can be favored with the changes in OLR and can also increase the response time even though acetate utilization by SRB occurred, since every degradation step delays the final sulfide concentration at steady state. The decrease of the HRT (tuning II) caused the acetate accumulation in the bioreactor (Table 5.1), while longer HRT allowed acetate utilization (Table 5.2). This is in agreement with Zhang and Noike (1994) who demonstrated that the HRT is a significant factor in the selection of the predominant microbial species, where acetogens were found to be particularly sensitive to changes in HRT.

The measurement of the acetate concentration as an additional output measurement is an option to adapt the PID controller for its application in bioreactors to control the sulfide concentration. This option has been applied by Alvarez-Ramirez et al. (2002) in anaerobic bioreactors to regulate the effluent COD concentration by using the volatile fatty acids concentration (including acetate) as a secondary measurement in the control gain configuration that is incorporated into the feedback loop scheme of a traditional PI controller to enhance the robustness of the control scheme with respect to influent disturbances. Another adaptation the PID parameters can be obtained from the information of the dynamics of the bioprocess such as reaction pathways and kinetics as well as mass balances, which are always required in biological systems due to their non linearity and non stationary characteristics (Steyer et al. 2000). This information can help to predict the response time of the system to the change applied and thus prevent excessive control action.

5.4.2 Tuning I strategy- advantages and drawbacks for sulfide control

The increase of the OLR following tuning I strategy (a reaction of the COD_{in}) showed rapid responses (Figure 5.4a), a high sulfide load change and almost negligible time delays (Table 5.3). However, COD overload and hence, substrate inhibition (Qatibi et al. 1990) as well as sulfide toxicity (O'Flaherty and Colleran 2000; Reis et al. 1992) may occur. Therefore, maximum permissible COD_{in} loads should be considered in the controller for obtaining the desirable sulfide concentration without risking loss of the biomass due to inhibition. This optimization in the controller has been previously reported in anaerobic process to avoid pH inhibition due to volatile fatty acids (Ryhiner et al. 1993).

Contrary to the important increment in the sulfide concentration when the change in the OLR is positive (Figure 5.4a), decreasing the OLR via tuning I displayed a slower response time (Figure 5.4a) and lower sulfide load changes (Table 5.3), and thus, a higher K_c (Table 5.3). This means that the sulfide production could be maintained under similar conditions despite the decrease in the COD_{in} concentration that lowers the substrate available. This was not related to an increase in the sulfate reduction rate (Table 5.2) nor a more efficient use of the substrate available as the COD removal efficiency prior to and after the step response for a positive and negative OLR change was similar (Table 5.2). It is also not due to a potential difference between the theoretical and real HRT that could delay the sulfide response to the change in OLR (Table 5.2 and Figure 5.3). Therefore, it is most probably related to the consumption of storage products accumulated during the COD increase. Indeed, SRB has the capacity to accumulate storage products under feast-famine conditions (Hai et al. 2004), which can be promoted with the increase and decrease of the OLR such as in this study. The step responses can originate periods of excess of carbon alternated with substrate limitation, favoring the selection of biomass with substrate storage capacity (Serafim et al. 2008). This possibility should thus be considered as it can affect the control of the sulfide concentration in bioreactors due to non-accounted substrate sources.

5.4.3 Tuning II strategy- advantages and drawbacks for sulfide control

An increment in the OLR via the tuning II strategy displayed a slower response time (Figure 5.4b) and less strong changes in the sulfide load as compared with tuning I (Table 5.3). Thus, tuning strategy II yielded to higher integral and derivative parameters (Table 5.3). This was due to the insufficient sulfate reducing activity (for lactate: 16.7 gCOD/gVSS*d, for acetate: 37.1 gCOD/gVSS*d) and SRB biomass (data not shown) to overcome the OLR increment applied. In addition, increasing the OLR via decreasing the HRT can yield to a sulfide and biomass washout (Kaksonen et al. 2004). In contrast, longer HRTs (a decrease in the OLR) caused accumulation of sulfide (Figure 5.4b and Table 5.2) as well as an increment in the COD and sulfate removal efficiencies (Table 5.3). Indeed, longer residence times allow for more SRB bioconversion, and thus, a reduction of the sulfide concentration is not observed (Figure 5.4b). Therefore, another strategy to decrease the sulfide concentration in the IFB bioreactor should be considered since both tuning strategies failed to achieve this (Figure 5.4). In addition, the PID controller parameters obtained for this action (Table 5.3) can become the bioreactor unstable due to the high values of the K_c that results in a large change in the output (Stephanopoulos 1984). One strategy to decrease the sulfide concentration is to dilute the effluent up to the required level for metal precipitation. This strategy is widely used as control action for obtaining the effluent quality criteria (Metcalf and Eddy 2002). Nevertheless, for this electron donor-deficient type of wastewaters (Papirio et al. 2012), this strategy would imply substrate losses, which contradicts one of the aims of sulfide control in bioreactors.

5.4.4 Validation of the pS electrode response

This study shows that the online measurement of the sulfide concentration via the pS can be used for the control of the sulfide concentration in SRB bioreactors. The use of the pS electrode in bioreactors gives an advantage over non-online methods for sulfide determination in bioreactors since the sulfide in the system is measured inside the bioreactor avoiding volatilization or oxidation (Hu et al. 2010) that causes a decrease of the reported sulfide values (Table 5.2), in contrast with the fairly stable pS reading values (Figure 5.4).

The pS electrode response is highly sensitive to pH variations (Figure 5.5b), as it influences the predicted values of the sulfur species (Beneš and Paulenová 1974; Al-Tarazi et al. 2004) and its voltage response is also less pronounced at high sulfide concentrations; (Figure 5.5a). The former is due to the logarithmic response of the pS values (Equation 1) to the changes in the TDS concentration (Figure 5.5b). Therefore, errors in TDS measurements above 100 mg/L are more likely to occur when the S^{2-} species is followed, as the solution becomes saturated in H_2S and therefore the maximum S^{2-} concentration is not attained (Petrucci and Moews 1962). Notwithstanding these characteristics, the pS electrode readings are valid for utilization in the control of SRB bioreactors as high sulfide concentrations in the bioreactor are not desirable since: 1) the H_2S can inhibit SRB (Chen et al. 2008; Kalyuzhnyi and Fedorovich 1998), 2) the necessary sulfide amount to precipitate metals for their recovery from wastewaters is rather low as metal concentrations in mine wastewaters range between 10 to 250 mg/L (Papirio et al. 2012). Thus, the sulfide concentrations remain below the pS sensitivity range limit.

5.5 Conclusions

- The sulfide concentration in bioreactors can be controlled by the variation of the COD_{in} concentration (high ΔTDS) or the HRT (low ΔTDS) depending on the desired evolution of the sulfide concentration. The PID parameters obtained via the manipulation of the COD_{in} concentration are adequate for the increase of the sulfide concentration. If the aim is to decrease the sulfide concentration, a dilution strategy can be applied for sulfide washout of the system.

- Delays in the response time and a high control gain were the most critical factors affecting the application of a control strategy in bioreactor that were caused by the induction of different metabolic pathways accounting sulfate reduction including the accumulation of storage products and substrate utilization.

- The pS electrode response is adequate for its application in controlling sulfate reducing bioreactors. However, pH variations and high sulfide concentrations should be carefully observed for correction of the pS values displayed.

References

Al-Tarazi M, Heesink ABM, Azzam MOJ, Yahya SA, Versteeg GF (2004) Crystallization kinetics of ZnS precipitation; an experimental study using the mixed-suspension-mixed-product-removal (MSMPR) method. Crystal Research and Technology 39 (8):675-685. doi:10.1002/crat.200310238

Alvarez-Ramirez J, Meraz M, Monroy O, Velasco A (2002) Feedback control design for an anaerobic digestion process. Journal of Chemical Technology & Biotechnology 77 (6):725-734. doi:10.1002/jctb.609

APHA APHA (2005) Standard methods for examination of water and wastewater. 20 edn., Washington D.C.

Beneš P, Paulenová M (1974) Surface charge and adsorption properties of polyethylene in aqueous solutions of inorganic electrolytes. Colloid & Polymer Science 252 (6):472-477

Bijmans MFM, Buisman CJN, Meulepas RJW, Lens PNL (2011) 6.34 - Sulfate Reduction for Inorganic Waste and Process Water Treatment. In: Editor-in-Chief: Murray M-Y (ed) Comprehensive Biotechnology (Second Edition). Academic Press, Burlington, pp 435-446

Cord-Ruwisch R (1985) A quick method for the determination of dissolved and precipitated sulfides in cultures of sulfate-reducing bacteria. Journal of Microbiological Methods 4 (1):33-36

Chen Y, Cheng JJ, Creamer KS (2008) Inhibition of anaerobic digestion process: A review. Bioresource Technology 99 (10):4044-4064. doi:10.1016/j.biortech.2007.01.057

Dan TB-B, Frevert T, Cavari B (1985) The sulfide electrode in bacterial studies. Water Research 19 (8):983-985. doi:10.1016/0043-1354(85)90366-5

Dunn IJ, Heinzle E, Ingham J, Prenosil JE (2003) Biological Reaction Engineering: Dynamic Modelling Fundamentals with Simulation Examples second edn. Wiley-VCH Verlag GmbH &Co. KgaA, Weinheim,

Dunn IJ, Heinzle E, Ingham J, Přenosil JE (2005) Automatic Bioprocess Control Fundamentals. In: Biological Reaction Engineering. Wiley-VCH Verlag GmbH & Co. KGaA, pp 161-179. doi:10.1002/3527603050.ch7

Gibert O, de Pablo J, Luis Cortina J, Ayora C (2004) Chemical characterisation of natural organic substrates for biological mitigation of acid mine drainage. Water Research 38 (19):4186-4196. doi:10.1016/j.watres.2004.06.023

Grootscholten T, Keesman K, Lens P (2008) Modelling and on-line estimation of zinc sulphide precipitation in a continuously stirred tank reactor. Separation and Purification Technology 63 (3):654-660

Gupta A, Flora JRV, Sayles GD, Suidan MT (1994) Methanogenesis and sulfate reduction in chemostats—II. Model development and verification. Water Research 28 (4):795-803. doi:10.1016/0043-1354(94)90086-8

Hai T, Lange D, Rabus R, Steinbüchel A (2004) Polyhydroxyalkanoate (PHA) accumulation in sulfate-reducing bacteria and identification of a class III PHA synthase (PhaEC) in Desulfococcus multivorans. Appl Environ Microbiol 70 (8):4440-4448

Heinzle E, Dunn I, Ryhiner G (1993) Modeling and control for anaerobic wastewater treatment

Bioprocess Design and Control. In, vol 48. Advances in Biochemical Engineering/Biotechnology. Springer Berlin / Heidelberg, pp 79-114. doi:10.1007/BFb0007197

Hu P, Jacobsen LT, Horton JG, Lewis RS (2010) Sulfide assessment in bioreactors with
gas replacement. Biochemical Engineering Journal 49 (3):429-434.
doi:10.1016/j.bej.2010.02.006

Jagadeesh CAP, Sudhaker RDS (2010) Modeling, Simulation and Control of
Bioreactors Process Parameters - Remote Experimentation Approach.
International Journal of Computer Applications 1 (10):81-88

Kaksonen AH, Franzmann PD, Puhakka JA (2004) Effects of hydraulic retention time
and sulfide toxicity on ethanol and acetate oxidation in sulfate-reducing metal-
precipitating fluidized-bed reactor. Biotechnology and Bioengineering 86
(3):332-343

Kaksonen AH, Puhakka JA (2007) Sulfate Reduction Based Bioprocesses for the
Treatment of Acid Mine Drainage and the Recovery of Metals. Engineering in
Life Sciences 7 (6):541-564

Kalyuzhnyi SV, Fedorovich VV (1998) Mathematical modelling of competition
between sulphate reduction and methanogenesis in anaerobic reactors.
Bioresource Technology 65 (3):227-242. doi:10.1016/s0960-8524(98)00019-4

König J, Keesman KJ, Veeken A, Lens PNL (2006) Dynamic Modelling and Process
Control of ZnS Precipitation. Separation Science and Technology 41 (6):1025 -
1042

Marsili-Libelli S, Beni S (1996) Shock load modelling in the anaerobic digestion
process. Ecological Modelling 84 (1–3):215-232. doi:10.1016/0304-
3800(94)00125-1

Metcalf, Eddy (2002) Wastewater Enginering: Treatmt & Reuse. McGraw-Hill
Education (India) Pvt Limited,

O'Flaherty V, Colleran E (2000) Sulfur problems in anaerobic digestion. In: Lens P,
Hulshoff Pol L (eds) Environmental Technologies to Treat Sulfur Pollution -
Principles and Engineering

IWA Press, London, UK, pp 467-489

Olivet D, Valls J, Gordillo MÀ, Freixó À, Sánchez A (2005) Application of residence
time distribution technique to the study of the hydrodynamic behaviour of a full-
scale wastewater treatment plant plug-flow bioreactor. Journal of Chemical
Technology & Biotechnology 80 (4):425-432. doi:10.1002/jctb.1201

Oyekola OO, Harrison STL, van Hille RP (2012) Effect of culture conditions on the
competitive interaction between lactate oxidizers and fermenters in a biological
sulfate reduction system. Bioresource Technology 104 (0):616-621.
doi:10.1016/j.biortech.2011.11.052

Oyekola OO, van Hille RP, Harrison STL (2009) Study of anaerobic lactate metabolism
under biosulphidogenic conditions. Water Research 43: 3345-3354

Papirio S, Villa-Gomez DK, Esposito G, Lens PNL, Pirozzi F (2012) Acid mine
drainage treatment in fluidized-bed bioreactors by sulfate-reducing bacteria: a
critical review. Critical reviews in environmental science and technology In
Press

Petrucci RH, Moews PC, Jr. (1962) H2S equilibria: The precipitation and solubilities of
metal sulfides. Journal of Chemical Education 39 (8):391

Pind P, Angelidaki I, Ahring B, Stamatelatou K, Lyberatos G (2003) Monitoring and
Control of Anaerobic Reactors

Biomethanation II. In: Ahring B, Ahring B, Angelidaki I et al. (eds), vol 82. Advances
in Biochemical Engineering/Biotechnology. Springer Berlin / Heidelberg, pp
135-182. doi:10.1007/3-540-45838-7_4

Qatibi AI, Bories A, J.L.Garcia (1990) Effects of sulfate on lactate and C2-, C3- volatile fatty acid anaerobic degradation by a mixed microbial culture. Antonie Van Leeuwenhoek 58 (4):241-249

Reis MAM, Almeida JS, Lemos PC, Carrondo MJT (1992) Effect of hydrogen sulfide on growth of sulfate reducing bacteria. Biotechnology and Bioengineering 40 (5):593-600. doi:10.1002/bit.260400506

Rodrigo MA, Seco A, Ferrer J, Penya-roja JM, Valverde JL (1999) Nonlinear control of an activated sludge aeration process: use of fuzzy techniques for tuning PID controllers. ISA Transactions 38 (3):231-241. doi:10.1016/s0019-0578(99)00018-x

Ryhiner GB, Heinzle E, Dunn IJ (1993) Modeling and simulation of anaerobic wastewater treatment and its application to control design: Case whey. Biotechnology Progress 9 (3):332-343. doi:10.1021/bp00021a013

Sampaio RMM, Timmers RA, Xu Y, Keesman KJ, Lens PNL (2009) Selective precipitation of Cu from Zn in a pS controlled continuously stirred tank reactor. Journal of Hazardous Materials 165 (1-3):256-265

Serafim L, Lemos P, Albuquerque ME, Reis MM (2008) Strategies for PHA production by mixed cultures and renewable waste materials. Applied Microbiology and Biotechnology 81 (4):615-628. doi:10.1007/s00253-008-1757-y

Stephanopoulos G (1984) Chemical process control. Prentice-Hall international series in the physical and chemical engineering sciences Prentice-Hall

Steyer J-P, Buffière P, Rolland D, Moletta R (1999) Advanced control of anaerobic digestion processes through disturbances monitoring. Water Research 33 (9):2059-2068. doi:http://dx.doi.org/10.1016/S0043-1354(98)00430-8

Steyer JP, Bernet N, Lens PNL, Moletta R (2000) Anaerobic treatment of sulfate rich wastewaters : process modeling and control. In: In: Environmental Technologies to Treat Sulfur Pollution / P.N.L. Lens and L.W. Hulshoff Pol. - London : IWA Publishing, 2000. - ISBN 1900222094. pp 207-235

Torner-Morales FJ, Buitrón G (2010) Kinetic characterization and modeling simplification of an anaerobic sulfate reducing batch process. Journal of Chemical Technology & Biotechnology 85 (4):453-459. doi:10.1002/jctb.2310

Veeken AHM, Akoto L, Hulshoff Pol LW, Weijma J (2003a) Control of the sulfide (S2-) concentration for optimal zinc removal by sulfide precipitation in a continuously stirred tank reactor. Water Research 37 (15):3709-3717

Veeken AHM, de Vries S, van der Mark A, Rulkens WH (2003b) Selective Precipitation of Heavy Metals as Controlled by a Sulfide-Selective Electrode. Separation Science and Technology 38 (1):1 - 19

Villa-Gomez D, Ababneh H, Papirio S, Rousseau DPL, Lens PNL (2011) Effect of sulfide concentration on the location of the metal precipitates in inversed fluidized bed reactors. Journal of Hazardous Materials 192 (1):200-207. doi:10.1016/j.jhazmat.2011.05.002

Villa-Gomez D, Enright AM, E. L, A. B, Kramer H, P.N.L. L (2012) Effect of hydraulic retention time on metal precipitation in sulfate reducing inverse fluidized bed reactors. Submited to Separation and Purification Technology

Warfvinge P (2009) Process Calculations and Reactor Calculations.

Widdel F (ed) (1988) Microbiology and ecology of sulfate- and sulfur-reducing bacteria. In The Biology of Anaerobic Microorganisms, .

Zagury GJ, Kulnieks VI, Neculita CM (2006) Characterization and reactivity assessment of organic substrates for sulphate-reducing bacteria in acid mine drainage treatment. Chemosphere 64 (6):944-954

CHAPTER
6

EFFECT OF PH ON
THE MORPHOLOGY, MINERALOGY AND SOLID-LIQUID
PHASE SEPARATION OF Cu AND Zn
PRECIPITATES PRODUCED WITH
BIOGENIC SULFIDE

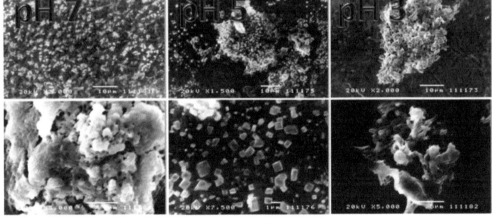

Abstract

The characteristics (morphology, mineralogy and solid-liquid phase separation) of the Cu and Zn precipitates formed with sulfide produced in a sulfate reducing bioreactor were studied at pH 3, 5 and 7. These characteristics were influenced by the dissolved organic matter and macronutrients present in the bioreactor liquid. The precipitates formed at pH 7 display faster settling rates, better dewaterability and higher concentrations of settleable solids as compared to the precipitates formed at pH 3 and 5. These differences were linked to the agglomeration of the sulfidic precipitates and co-precipitation of the phosphate added to the bioreactor influent. The Cu and Zn quenched the intensity of the dissolved organic matter peaks identified by fluorescence-excitation emission matrix spectroscopy suggesting a binding mechanism that decreases supersaturation, especially at pH 5. X-ray absorption fine structure spectroscopy analyses confirmed the precipitation of Zn-S as sphalerite and Cu-S as covellite in all samples, but also revealed the presence of Zn-sorbed on hydroxyapatite. These analyses further showed that CuS structures remained amorphous regardless the pH, whereas the ZnS structure was more organized at pH 5, as compared to the ZnS formed at pH 3 and 7, in agreement with the cubic sphalerite type structures observed through scanning electron microscopy at pH 5.

This Chapter was submitted for publication as:

Villa-Gomez D.K., van Hullebusch E.D., Maestro R., Farges F., Nikitenko S., Kramer H., Gonzalez-Gil G. and Lens P.N.L. (2013) Morphology, mineralogy and solid-liquid phase separation characteristics of Cu and Zn precipitates produced with biogenic sulfide. Submitted to Environmental Science and technology.

6.1 Introduction

Mining activities all over the world result in pollution problems due to the emissions of acid mine drainage (AMD) into the environment (Gazea et al. 1996; Johnson and Hallberg 2005). AMD results from the transformation of pyrite and other sulfide minerals into a leachate containing dissolved metals, sulfate and acidity upon their exposure to oxygen and water in the presence of sulfur oxidizing bacteria. The water pH progressively decreases, resulting in the mobilization of metals from the mine wastes to water streams (Neculita et al. 2007). These metal containing waste streams need to be treated for environmental reasons but also for the recovery of valuable metals.

Recovery of metals as metal sulfide precipitates is an appealing alternative to the conventional treatment methods such as hydroxide precipitation for several reasons: 1) the formation of highly insoluble salts even at pH values below 7 (Brooks 1991); 2) possibility of selective metal recovery due to the differences in solubility product of the various metal sulfides over the pH range (Sampaio et al. 2009); 3) better settling, thickening and dewatering characteristics compared to hydroxide-treated sludge (Lewis and Swartbooi 2006; Esposito et al. 2006; Djedidi et al. 2009) and 4) direct reuse of Cu and Zn precipitates as many metal refining operations are designed for processing these metal sulfide ores (Brooks 1991). Despite these advantages, the use of metal sulfide precipitation for metal recovery is still limited by the challenging solid-liquid phase separation, since the predominance of nucleation over crystallization favours the formation of small precipitates known as fines that are difficult to settle (Lewis 2010).

Biogenic sulfide produced by sulfate reducing bacteria (SRB) is an alternative to chemically produced sulfide to precipitate and recover metals from AMD. The sulfate present in most AMD streams is reduced to sulfide by SRB when an organic electron donor or hydrogen is added (Kaksonen et al. 2003; Esposito et al. 2006; Gallegos-Garcia et al. 2009; Kaksonen and Puhakka 2007). Investigations on metal precipitation using biogenic sulfide have mainly focused on metal removal efficiencies, with little attention given to the morphology and mineralogy of the metal sulfide precipitates that determine the ability to separate the precipitates from the wastewaters. Compounds commonly present in the biogenic sulfide such as organics and nutrients supplied for bacterial growth affect the size and structure of the metal sulfide precipitates (Esposito et al. 2006; Sampaio et al. 2009; Bijmans et al. 2009). Knowledge regarding the influence of these key components on the metal precipitate characteristics such as particle size and solid-liquid separation is important to optimize the metal recovery efficiency.

The objective of this study was thus to determine the characteristics of the metal (Cu and Zn) precipitates formed with biogenic sulfide. Metal precipitation experiments using biogenic sulfide were performed for the evaluation of the efficiencies in solids-liquid separation trough particle size, settling rate and dewaterability analysis at different pH values. Mineralogical analysis and dissolved organic matter characterization were carried out to correlate the results with the conditions tested. X-ray absorption fine structure (XAFS) spectroscopy was used as a probe of the local atomic structure of the metal sulfides since they crystallize poorly and/or are found immerse into biological products (Lenz et al. 2011; Prange and Modrow 2002) and therefore, species-specific analytical methods are required.

6.2 Materials and methods

6.2.1 Metal precipitation experiments

Three sets of metal precipitation experiments (A, B, and C) using biogenic sulfide were tested at three different pH values (3, 5 and 7) keeping the metal to sulfide ratio constant at stoichiometric values with only Zn (Experiment A), under excess of metals (Experiment B) and marginally under excess of sulfide with Zn and Cu (Table 6.1). The experiments were performed at room temperature (23 ± 2 °C) in serum bottles of 117 mL containing 100 mL of biogenic sulfide and 17 mL of a Zn (A) or Cu-Zn (B and C) concentrated solution to obtain the final metal concentration desired in the experiment (Table 6.1). Once the biogenic sulfide was poured into the bottles, the headspace was bubbled with nitrogen gas for 5 to 10 seconds to strip out the air prior to adding the metal solution to avoid oxidation. The pH was adjusted by adding the required amounts of 1.0 M HCl or 1.0 M NaOH. After the pH adjustment, the sulfide concentration was measured and the metal solution added. Experiments were done in triplicate. The lactate, acetate and sulfate concentration of the biogenic sulfide used on each experiment is also shown in Table 6.1.

The biogenic sulfide was taken from an inverse fluidized bed (IFB) reactor described by Villa-Gomez et al. (2011), operating at an organic loading rate of 2 gCOD/L*day with a COD/SO_4^{2-} ratio of 0.67 (mol/mol) and at an HRT of 24 h. Table 6.1 the IFB reactor influent composition.

Table 6.1 Composition of the biogenic sulfide used in the batch experiments and composition of the bioreactor influent used for biogenic sulfide production.

Composition of the biogenic sulfide (mM):				Composition of the bioreactor influent (mM):	
Experiment	A	B	C	KH_2PO_4	1.47
Sulfide [HS^-, S^{2-}]	5	7.2	1.1	NH_4Cl	3.74
[Zn^{2+}]	4.7	6.8	0.5	$CaCl_2 \cdot 2H_2O$	0.102
[Cu^{2+}]	0	6.8	0.4	$MgCl_2 \cdot 6H_2O$	0.59
[$\sum M^{2+}$]/[S^{2-}] ratio	1	2	0.8	KCl	3.35
Acetate	4.9	13.1	*	$C_3H_5NaO_3$	20.8
Lactate	10.7	5.3	15.6	Na_2SO_4	31.2
Sulfate [SO_4^{2-}]	25	20.3	13.7		

$\sum M^{2+}$: Metals, *: Below detection limit.

6.2.2 Analyses

Metal measurements in the liquid phase were done three hours after the start of the experiment (A, B and C) to determine metal removal efficiencies from the liquid phase as a consequence of a reaction with the biogenic sulfide. Flame spectroscopy (AAS Perkin Elmer 3110) and furnace spectroscopy (AAS Solaar MQZe GF95) was used after dilution, filtration (0.45 μm pore size) and acidification of the samples with concentrated HNO_3. Dissolved sulfide was determined spectrophotometrically by the colorimetric method described by Cord-Ruwisch (1985) using a Perkin Elmer Lambda20 spectrophotometer.

6.2.3 Analyses to the precipitates from experiment C

Total suspended solids (TSS), non-volatile suspended solids (NVSS) and volatile suspended solids (VSS) concentration were determined according to standard methods (APHA 2005). The dewatering properties of the precipitates were assessed by the capillary suction time (CST) using a CST test kit (Triton CST Filterability Tester, model 200, Triton Electronics Ltd., Essex, UK) using standard filter papers and a 18 mm sludge reservoir. The CST value recorded was standardized to a TSS concentration of 1 g/L.

Particle size distribution (PSD) analysis were done following the procedure from Villa-Gomez et al. (Villa-Gomez et al. 2012) using a particle size analyzer (Microtrac 53500). The PSD varied greatly after each instrument run with the same sample, therefore, sonication was applied to remove macroscale agglomeration as described by Villa-Gomez et al. (Villa-

Gomez et al. 2012). The mean particle size (D_{50}) of each sample was calculated after determining the minimum and maximum sizes contributing to the highest peaks.

Scanning electron microscopy (SEM) was performed on a Jeol JSM-5400 for visual characterization of the precipitates at 1, 10 and 100 μm resolution. The precipitates were collected with a plastic fine bore Pasteur pipette and placed on an aluminum pin and sputtered with carbon.

A small amount of precipitates was collected and centrifuged at 5000 rpm during 10 minutes for mineralogical characterization using XAFS experiments, including X-ray absorption near edge spectroscopy (XANES) and extended X-ray absorption fine structure (EXAFS) regions. The spectra were collected on the DUBBLE beam line BM26A of the European Synchrotron Radiation Facility (Grenoble, France) (Borsboom et al. 1998). Spectra were also collected from Cu and Zn reference compounds, including ZnS (sphalerite), Zn-acetate and Cu-acetate dihydrate. Zn sorbed on apatite and $Zn_3(PO_4)_2 \cdot 4H_2O$ (hopeite) were taken from previous experiments of Villa-Gomez et al. (2011). CuS (covellite) and Cu_2S (chalcocite) spectra were taken from the literature (Parsons et al. 2002; Pattrick et al. 1997) and the EXAFS/XANES database of standard materials (http://x18b.nsls.bnl.gov/data.htm).

The X-ray energy was varied from 200 eV below to 750 eV above the absorption K-edge of Zn (9659 eV) and Cu (8984 eV). The sample analysis and data acquisition followed the same procedure as Villa-Gomez et al. (Villa-Gomez et al. 2012). Zn spectra were first collected followed by Cu spectra. It was observed that the X-ray beam induced photoreduction of Cu in the samples, as confirmed by the growing presence of native copper in overexposed samples. To avoid this artefact, beam exposure was minimized as much as possible. Also, each sample was divided into two parts for EXAFS data collection at the Zn and at the Cu K-edge, respectively. In addition, the X-ray beam was defocused while data was collected in a cryostat (liquid nitrogen) operated at 80 K.

EXAFS spectra were normalized using the XAFS software package (Winterer 1997) using standard procedures (Farges et al. 2001). A wavelet analysis of the EXAFS spectra was performed according to Muñoz et al. (Muñoz et al. 2003) to better define the various atomic pairs contributing to the spectroscopic data at the Zn and Cu K-edges. The goal of those computations is to display the normalized EXAFS spectrum for each sample (on the X-axis), associated to its computed Fourier Transform, FT (on the Y axis) with, in between those last two spectra, the computed wavelet diagram. The latter displays 3 dimensions: k-space (like the EXAFS), (R+Δ) space (like the FT) and the intensity of the wavelet indicated as a continuum of greys.

Experiment C was repeated to determine the settling rate of the precipitates by transferring the whole sample (liquid+solid phase) to Imhoff settling graduated cones. Settleable solids were calculated following standard procedures (APHA 2005). The height of the liquid/sludge interface was registered every 30 minutes up to 2 h. The results are presented as h/h_0 vs time, where h is the height of the settled volume at time $t_{(h)}$ and h_0 is the height of the settled volume at time = 0 h.

6.2.4 Fluorescence-Excitation Emission Matrix (F-EEM) spectroscopy analysis

F-EEM spectroscopy analysis was used to characterize the dissolved organic matter (DOM) changes with the pH and metal addition following the experimental conditions of experiment C. The DOM concentration was determined as dissolved organic carbon (DOC) by a total organic carbon analyzer (TOC-VCPN (TN), Shimadzu, Japan) on filtrated samples (0.45 μm pore size) to adjust the samples to 1 mg/L of DOC before F-EEM spectroscopy analysis. F-EEM spectra were then obtained using a FluoroMax-3 spectrofluorometer (HORIBA Jobin Yvon, Edison, NJ, USA) following the procedure described in Maeng et al. (Maeng et al. 2011). The 3-D maps were built using Matlab 7.5 software. Coordinates of the main noticeable peaks were established in these maps.

6.2.5 Speciation modeling

Visual MINTEQ version 3.0 (US EPA, 1999) was used to calculate speciation, saturation indices and equilibrium of solid and dissolved phases in all the experiments (A, B and C). The influent composition used in the bioreactor and the effluent lactate, acetate and sulfate concentration present in the biogenic sulfide (Table 6.1) were used as input in the model. The pH was fixed while for each run, the model calculated the ionic strength. The program was set to give information of the precipitation efficiencies and mineral identity of the precipitates formed at equilibrium as well as the saturation index of the theoretical precipitates formed (oversaturated species) prior to (predecessor) and at equilibrium. The saturation index of the oversaturated species refers to the logarithm of the ion activity product minus the logarithm of the solubility product.

6.3 Results

6.3.1 Metal removal/precipitation efficiencies

Figure 6.1a shows the Zn removal efficiency in the batch experiments versus the predicted precipitation efficiency at equilibrium obtained with the speciation model. The Cu removal efficiency in the precipitation experiments was always high (>88.1%) as in the calculated model despite the differences in pH and biogenic sulfide compounds concentration. Therefore, the Cu removal efficiencies in the experiments are only shown in the supporting information. In contrast, the Zn removal efficiency as a function of pH showed considerable fluctuation, ranging from 5.9 to 94.6% (Figure 6.1a). In the experiments A, in which no Cu was present, the Zn removal efficiency was low despite the sufficient stoichiometric sulfide concentration for ZnS precipitation (approximately 6% at pH 3 and 50% at pH 5 and 7). In the experiments B, in which both Zn and Cu were present in equimolar quantities, the Zn removal efficiency was 7.4, 93 and 94.6% at pH 3, 5 and 7, respectively. At pH 3, the measured removal efficiency (7.4% ±3) fairly agreed with the precipitation efficiency calculated using the speciation model (Figure 6.1). In contrast, at pH 5 and pH 7, measured values were 2 to 3 times higher than the predicted precipitation efficiency by the speciation model, suggesting that additional mechanisms than those considered in the model were involved in Zn removal. For the experiments C, the Zn removal efficiency was high for pH 3 and 7 as predicted in the speciation model, while for pH 5 it was notably lower (69.5% ±13.4).

The speciation model shows that oversaturated species, and therefore possible precipitates, were mainly due to sulfide binding (Figure 6.1b). The main predecessor of the Cu precipitates was covellite (Cu(I)S), and for Zn, sphalerite (ZnS) and wurtzite (ZnS). At pH 7, a further diversity of predecessor-precipitated species were found, including phosphate precipitates as $Zn_3(PO_4)_2*4H_2O$ and hydroxyapatite $(Ca_5(PO_4)_3(OH))$ (Figure 6.1b).

A proportional increase of the saturation index is observed with the increase of the pH for covellite and sphalerite, nevertheless, covellite values are much higher (Figure 6.1b). The organic compounds from the biogenic sulfide: lactate and acetate, remained in the dissolved state without being involved in the precipitation of neither Zn nor Cu (data not shown). The species predicted at the final thermodynamic equilibrium were sphalerite for Zn and covellite for Cu in all experiments (data not shown).

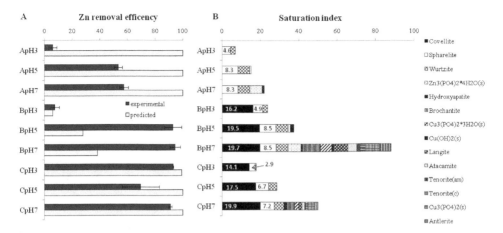

Figure 6.1 A) Zn removal efficiencies in the batch experiments and their standard deviations versus the predicted precipitation efficiencies at equilibrium in the speciation model and B) saturation index of the oversaturated species predicted in the speciation model. Data labels are given for the saturation index of covellite and sphalerite.

6.3.2 Metal precipitates characterization

6.3.2.1 Particle size and morphology

The measured PSD of the samples showed multiple D_{50} peaks. Figure 6.2 shows the volume fractions of these peaks versus their position in the PSD. At the pH values tested, the majority of the peaks fall in the range of 0.5 to 7 µm, while larger D_{50} peaks are observed at pH 3 and 5 that are probably loosely bound aggregates. At pH 7, three peaks were identified around 0.5, 3 and 7 µm. A more heterogeneous precipitates with respect to size were observed at pH 3 and 5 with peak values ranging from 5.6 to 35.6 µm and from 0.5 to 132 µm, (data not shown) respectively.

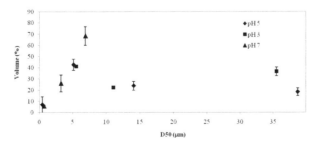

Figure 6.2 Volume fractions of the identified peaks in the PSD of the precipitates formed at pH 3, 5 and 7 in experiments C.

The SEM images of the solids collected at the three pH values tested showed aggregation of particles with no crystal forms observable at pH 7 (Figure 6.3a). At pH 5, a large number of small cubic crystals over a wide range of sizes below 10 μm are observed, but no aggregation (Figure 6.3b). At pH 3, a different aggregation characteristic is observed as compared with the SEM images of the precipitates formed at pH 7, where the precipitates agglomerated as clumps with smooth surface (Figure 6.3c).

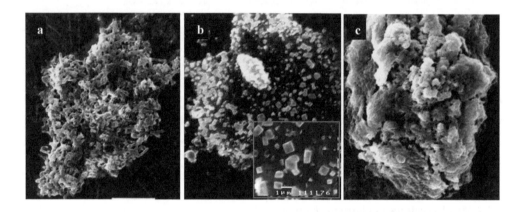

Figure 6.3 SEM pictures of the precipitates formed at pH: a) 7, b) 5 and c) 3 at 10 μm magnification (zoomed picture B at 1 μm magnification).

6.3.2.2 Mineralogical analysis of the Zn and Cu precipitates

The Zn K-edge XANES spectra for the precipitates at the three pH values tested suggest the presence of Zn-S compound as sphalerite (Figure 6.4). The spectroscopic similarities with sphalerite are highest at pH 5, where XANES features A to E are almost identical, while at pH 3, these features are in a lesser extent. In contrast, at pH 7, features A to E are much broader as compared to sphalerite, thus suggesting the presence of a much less ordered local structure in the Zn-S pairs. The feature A is remarkably different, indicating a contribution from another Zn environment (Figure 6.4). At the same pH, the feature E is slightly shifted towards lower energies, suggesting the presence, in that sample, of minor amounts of Zn-O environments, as in Zn-sorbed hydroxyapatite, $Zn:Ca_5(PO_4)_3 \cdot (OH)$ (Figure 6.5).

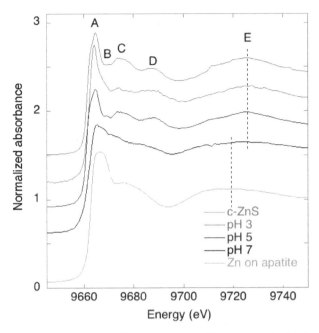

Figure 6.4 Zn K-edge XANES spectra for selected model compounds (ZnS sphalerite and Zn sorbed on apatite) as compared to the samples of the precipitates formed at pH 3, 5 and 7. Capital and small letters indicate the main features of the reference compounds.

The wavelet diagrams computed for the various Zn K-edge EXAFS spectra collected for this study as well as for the ZnS model compound (Figure 6.5) show, accordingly, first neighbours S as an elongated spot centered near 7 Å^{-1} (Muñoz et al. 2003). Hence, Zn-S pairs are identified in all batch samples, as in sphalerite (Figure 6.4 and 6.5). Due to their relatively higher backscattering amplitude, second neighbours Zn are characterized by a spot centered at slightly higher k-values, namely around 8 Å^{-1} (Muñoz et al. 2003). These next-nearest Zn neighbours are only present for pH 7 and 5, while for pH 3, those contributions are, if present, too weak and/or disordered to be detected by EXAFS spectroscopy (Figure 6.5) thus suggesting that these precipitates are somewhat amorphous. In contrast, ZnS at pH 5 shows the presence of next-nearest Zn neighbors and thus, a more crystallized structure, where the presence of significant amounts of colloidal and/or aperiodic and/or amorphous phase is not observed. Contrary to the suggestions on the Zn XANES spectra (Figure 6.4), no significant contribution from O first neighbours around Zn was observed at pH 7. This is due to the fact that those Zn-O pairs must be too highly disordered to be detectable in the EXAFS region, whereas disorder effects are much less destructive in the XANES region (Farges 2001).

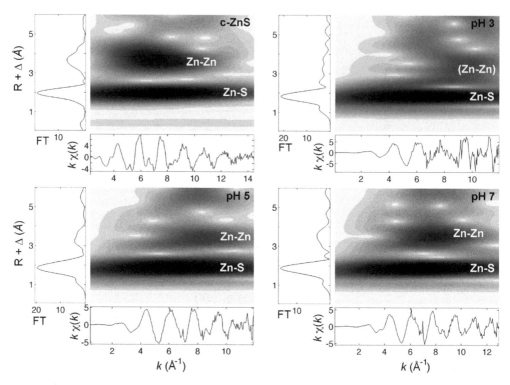

Figure 6.5 Wavelet analyses of the normalized Zn K-edge EXAFS spectra for sphalerite (top left) as compared to the samples of the precipitates formed at pH 3, 5 and 7. The spots are labelled for the corresponding atomic pairs. The wavelet modulus ranges from zero (white) to its maximum value (black).

The XANES spectra at the Cu-K edge for all the samples (Figure 6.6) showed close to that for covellite when compared the feature A-F from the reference compounds ($Cu(I)_2S$ (covellite) and $Cu(I)S$ (chalcocite), while the pre-edge features related to $Cu(0)$ (foil) and $Cu(II)$ (ac) (labelled a, b and c on Figure 6.6; (Farges et al. 2006)) are not visible in the three batch samples. In particular, the XANES spectrum for the pH 5 sample is the closest to covellite. Despite Cu is also monovalent and coordinated by sulfide first neighbours, the speciation of Cu in the samples at pH 3 and 7 is slightly different from pH 5, probably as a result of radial distortion (Farges et al. 2006).

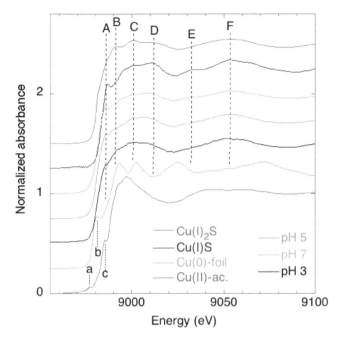

Figure 6.6 Cu K-edge XANES spectra for selected model compounds containing either Cu(0)(as native copper), Cu(I) (as sulfides) and Cu(II) (as an organic-oxide) as compared to the samples of the precipitates formed at pH 3, 5 and 7. Capital and small letters indicate the main features of the reference compounds.

Figure 6.7 shows the wavelet diagrams computed for the various Cu K-edge EXAFS spectra collected for this study as well as for models compounds of Cu(0), Cu(I) and Cu(II). The wavelet diagrams confirm that the Cu-containing precipitates at all pH values are made primarily of Cu(I)-sulfides, with little or no forms of Cu(0) nor Cu(II). The lack of significant amounts of native Cu(0) confirms the absence of photoreduction artefacts that could have affected our measurements. The lack of Cu(0) and Cu(II) is also consistent with the absence of Cu and O first neighbours, in inorganic or organic form, around Cu in the samples. As for the XANES spectra, the analogies with covellite are the highest, with Cu-S pairs at 1.9 Å on the (R+Δ) axis and Cu-Cu pairs between 3 and 4 Å. Those uncorrected distances correspond to ~ 2.2 and 3.2-3.8 Å for Cu-S and Cu-Cu pairs in covellite (Goh et al. 2006). Therefore, the predominant Cu speciation in all samples is covellite despite that the Cu-Cu pairs appear less intense in the precipitates of this study (Figure 6.7), thus suggesting a more disordered local structure (amorphous) around Cu.

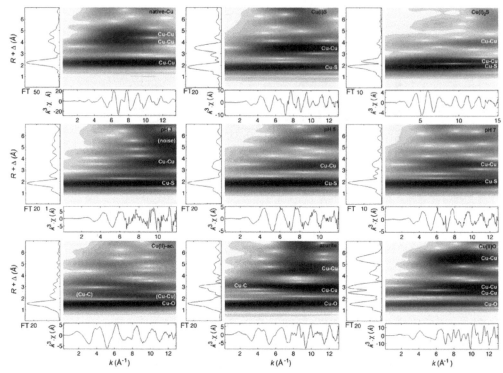

Figure 6.7 Wavelet analyses of the normalized Cu K-edge EXAFS spectra. The top three are reduced forms of Cu, namely native copper (left), covellite (middle) and chalcocite (right). The middle section show the samples of the precipitates at pH 3, 5 and 7. The lower three wavelets correspond to oxidized compounds of Cu, namely Cu-acetate dihydrate (left), azurite, a copper hydroxycarbonate (middle) and tenorite, CuO (right). The various Cu-X pairs (where X=O, S, Cu, C) are shown. The wavelet modulus ranges from zero (white) to its maximum value (black).

6.3.3 Influence of pH on the solid-liquid phase separation

The solid-liquid phase separation characteristics of the metal precipitates were pH-dependent (Figure 6.8). The settling rate after metal addition showed a similar decrease rate behavior in the dimensionless height of the solid/liquid interface but higher settling velocity for the precipitates formed at pH 7, followed by an increase in height at the end of the experiment due to the thickening of the sludge (Figure 6.8a).

The metal precipitates formed at pH 7, showed a faster settling rate (Figure 6.8a), more settleable solids and better dewaterability (Figure 6.8b) as compared to the other pH values tested. The increase in pH from 3 to 7 increased the settleable solids from 5.1 to 9.4 mL/L and decreased the CST from 59 to 33 ms. The settleable solids were composed of inorganic solids (NVSS), most probably due to the metal precipitation, but also contained a considerable amount of organic matter as observed in the VSS (Figure 6.8b), especially at pH 5 (128 mg/L). At pH 7, the precipitates contained a higher amount of NVSS (217.7 mg/L) as compared to the other precipitates that could be linked to a higher amount of precipitated species as predicted by the speciation model (Figure 6.1b).

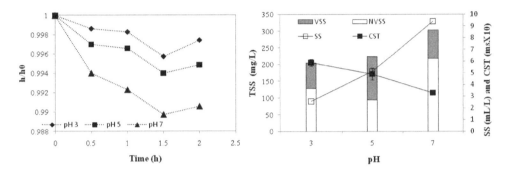

Figure 6.8 A) Settling rate and B) dewaterability (including VSS and NVSS concentrations) of the precipitates produced at pH 3, 5 and 7 in experiment C.

6.3.4 Influence of pH on metal quenching by fluorescencent dissolved organic matter

Four main peaks were identified by the F-EEM spectroscopy analysis in the biogenic sulfide suspensions prior to the metal addition (A, B, C and D) with higher intensities at pH 7 (Table 6.2). The peaks were detected at excitation/emission wavelengths of 270-280 nm/312-354 nm (A), 270 nm/298-336 nm (B), 370-410 nm/450-454 nm (C) and 270 nm/452-454 nm (D). The identified peaks were located in the humic (peak C) and fulvic (peak D) acid-like region and in the soluble microbial byproducts-like region (A and B) (Chen et al. 2003). Peak A and B location from this study have been identified for extracellular polymeric substances (EPS) in anaerobic methanogenic sludge (Sheng and Yu 2006; Ni et al. 2009) as well as in anaerobic sludge sulfate/ethanol rich wastewater (Bhatia et al. 2013).

Prior to metal addition to the biogenic sulfide suspensions, peaks A and B had similar intensities at all the pH values tested, whereas the intensity of the peaks C and D decreased markedly as the pH level decreased (Table 6.2). The addition of metals to the biogenic sulfide suspensions induced a notable decrease in the fluorescence signal of all the identified peaks, with different intensities depending of the pH value. In addition, peak A location slightly

135

shifted in all samples and new peaks were observed near peaks A and B for all pH values as well as for pH 7, respectively. The intensity of the peaks C and D highly decreased after metal addition for pH 3 and 5, while for pH 7, the quench was much weaker. The strongest quenched of the peaks intensity after metal addition was observed at pH 5 for all peaks (Table 6.2).

Table 6.2 Fluorescence spectral results of the samples at various pH values without and with metal addition.

	Peak A			Peak B			Peak C			Peak D		
	Em	Ex	Int	Em	Ex	Int	Em	Ex	Int	Em	Ex	Int
Samples prior metal addition												
pH 3	312	280	131.75	298	270	122.75	454	400	2198.4	452	270	1830.9
pH 5	342	280	146.84	298	270	137.06	454	400	2086	452	270	2397.2
pH 7	312	280	136.18	336	280	147.04	454	400	3781.6	452	270	2914.8
Samples after metal addition												
pH 3	336	270	118.39	324	270	111.12	450	410	506.55			
	344	280	119.35									
pH 5	340	280	80.1	406	270	79.18	452	370	43.42	454	270	60.44
	312	280	69.01									
pH 7	308	280	83.53	298	270	97.86	454	410	2395.9	454	270	1992.8
	354	280	67.52	324	270	69.19						

Em: emission; Ex: excitation and Int: Intensity

6.4 Discussion

6.4.1 Metal removal and solid-liquid phase separation of the metal sulfide precipitates

This study shows that the bioreactor influent and the DOM produced in the bioreactor affected the metal removal efficiencies and solid-liquid phase separation characteristics of the metal sulfide precipitates, and their effect was pH dependent. Zn and Cu precipitation was mostly ascribed to a reaction with the sulfide species (Figure 6.1b, 3-6). However, components present in the biogenic sulfide modified the structure (Figure 6.2 and 6.3), purity (Figure 6.1b, 6.4-6.7) and agglomeration of these precipitates, thus affecting their solid-liquid phase separation characteristics (Figure 6.8).

The variations in the metal removal efficiencies (Figure 6.1a) were related to the particle size of the metal sulfides. Particle sizes below 0.45 μm, which could not be retained with the filter used to determine metal removal efficiencies, were only observed at pH 5 on the SEM images (Figure 6.3b), while at pH 7, the images clearly show agglomeration of the preciptiates that increases the particle size (Mersmann 1999; Villa-Gomez et al. 2011). Accordingly, the formation of small precipitates with no agglomeration (Figure 6.3c) explains the low removal efficiency observed at pH 5 (Figure 6.1a). Parallel, even though the CuS precipitates were amorphous (Figure 6.7) and small precipitates (fines) are known to be formed (Lewis 2010), high Cu removal efficiencies at all pH values investigated were achieved (Figure 6.1a) due to the tendency of the CuS precipitates to agglomerate (Mokone et al. 2010; Sahinkaya et al. 2009), as their surface is charged (Al-Tarazi et al. 2005).

The tendency of the particles to agglomerate occurs due to a reduction of the electrical forces of repulsion among particles as in the electro/coagulation process (Lai and Lin 2004) and, in the case of metal sulfide precipitates, the supersaturation level plays an important role as nucleation promotes agglomeration (Al-Tarazi et al. 2004; Mokone et al. 2010). The speciation model predicted a proportional increase of the saturation index with the increase in pH (Figure 6.1a), thus, less agglomeration and consequently lower removal efficiencies are expected towards low pH values, at which the supersaturation diminishes (Peters et al. 1984). However, this argument contradicts the high Zn removal efficiencies encountered at pH 3 (Figure 6.1a), and thus, factors such as the DOM content from the biogenic sulfide, contributed the agglomeration of the metal sulfide precipitates as well.

Agglomeration, together with the particle size, structure and composition of the metal sulfide precipitates, contributes to the solid-liquid phase separation characteristics [46]. Contrary with previous authors (Mokone et al. 2010; 2003), faster settling rates (Figure 6.8a), shorter CST and higher concentrations of settleable solids of the metal sulfide precipitates (Figure 6.8b) were obtained at higher pH values (pH 7), regardless the differences in particle size (Figure 6.2). Veeken et al. (2003) calculated the settling velocities of ZnS particles and found that particles with a D_{50} of 3 μm d isplayed a much lower settling rate (0.006 m/h) as compared with particles with a D_{50} of 10 μm (0.6 m/h). Overall results thus suggest that the metal removal efficiencies and the solid-liquid phase separation results were influenced by the DOM and the macronutrients present in the bioreactor medium, which were not present in the aforementioned studies. The hydroxyapatite precipitated at pH 7, for instance, could have increased the precipitate concentration and consequently, the NVSS content and settleable solids concentration (Figure 6.8b) and could have also influenced the dewatering characteristics (Figure 6.8b). However, since hydroxyapatite formation is pH dependent (as referred in the results from the Visual Minteq model), their contribution to the solid-liquid phase separation characteristics is expected to be rather small upon the decrease of the pH such as in AMD waste streams.

4.2 Effect of the DOM

This study shows that the DOM present in the biogenic sulfide (Table 6.2) contributed to the characteristics of the metal sulfide precipitates, and their influence was pH dependent. A higher fluorescence intensity of the peaks (previously reported as EPS (Chen et al. 2003; Sheng and Yu 2006; Ni et al. 2009; Bhatia et al. 2013)) was observed at pH 7, as compared to pH 5 and 3 (Table 6.2). This intensity barely decreased after metal addition at pH 7 (Table 6.2), demonstrating that the molecular structure of the DOM was not modified with the metal addition. Despite this, the DOM could have been involved in the agglomeration of the precipitates as observed in the SEM pictures (Figure 6.3) and the dewatering characteristics (Figure 6.8). In natural environments, extracellular proteins are reported to promote the precipitation and agglomeration of ZnS precipitates (Moreau et al. 2004; Moreau et al. 2007). Additionally, EPS are known to induce the formation of bioflocs in activated (Bala Subramanian et al. 2010) and anaerobic granular (D'Abzac et al. 2010) sludge, thus contributing to their structural, surface charge and settling properties as well as dewaterability (Bala Subramanian et al. 2010; Raynaud et al. 2012) and this contribution is pH dependent (Chen et al. 2001; Sobeck and Higgins 2002).

In contrast with the results at pH 7, the fluorescence intensity highly deceased at pH 5 after metal addition (Table 6.2), coinciding with a higher amount of VSS in the settled sludge (Figure 6.8b). Therefore, it is inferred that the metals bound to the DOM at pH 5. The binding of metals with DOM have been suggested to decrease the supersaturation of the metal sulfide precipitation and hence, to allow crystallization over nucleation (Gadd 2009; Johnson et al. 2007). In the case of Zn, this corroborates with the crystalline structures observed in the SEM pictures (Figure 6.3) and the more organized sphalerite structure observed in the EXAFS analysis at pH 5 (Figure 6.5). In the absence of another reactive agent such as sulfide, the increase of the pH makes the metals-DOM interaction stronger. This is due to a variation in charge density of the organic molecules (Piccolo 2002), since the functional groups such as -COOH and -OH become available for binding with metals. However, metal complexation with DOM in the presence of sulfide is rather different as sulfide also complexes with the DOM functional groups (Rozan et al. 2000).

4.3 Solid phase speciation and morphology assessed by EXAFS and XANES spectroscopy

The XANES and EXAFS analyses showed the speciation of the Zn and Cu precipitates formed with biogenic sulfide, which is difficult to observe by commonly used techniques such as microscopy and X-ray diffraction analysis (Neculita et al. 2007; Villa-Gomez et al. 2012b; Brown and Calas 2012). Covellite was the only Cu precipitate formed at all the pH values tested, which was formed with Cu (I) (Figure 6.6), as previously reported on CuS formation under biotic (Ehrlich et al. 2004) and abiotic (Goh et al. 2006; Pattrick et al. 1997; Ehrlich et al. 2004) conditions. The EXAFS analysis also suggested a disordered local, and thus, less crystalline structure in all samples (Figure 6.7). This may be caused by the instantaneous reaction kinetics and the low CuS solubility over the entire pH range (Lewis 2010), which makes CuS precipitation predominant over the formation of another Cu species but also leads, as a consequence, to the formation of amorphous CuS precipitates (Figure 6.7). The latter confirms the findings of Shea and Helz (Shea and Helz 1989), who demonstrated that by mixing $CuCl_2$ with NaHS under acidic conditions, poorly crystalline CuS precipitates were formed. The pH can, nevertheless, affect CuS crystallinity under abiotic conditions as demonstrated by Rickard (Rickard 1972), who found that both covellite and "blaubleibender covellite" were crystalline at high pH values (pH>7). However, these had been subjected to considerable (several days) ageing and heating to enhance the crystallization process.

Whereas Cu undeniably precipitates only with sulfide and both the XANES and EXAFS analyses confirmed the presence of Zn-S as sphalerite in all samples analyzed (Figure 6.4), the EXAFS analysis revealed another Zn environment for the samples of the precipitates at pH 7 (Figure 6.5). Figure 4 shows the presence of minor amounts of Zn-O environments, as in Zn-sorbed hydroxyapatite. Thus, the predicted presence of hydroxyapatite at pH 7 is plausibly responsible for the difference in Zn speciation observed at this pH. The contribution of hydroxyapatite on the removal of Zn is not observed at pH 3 and 5, because the formation of this compound is rather pH dependent. A previous study (Villa-Gomez et al. 2012b) also confirmed that Zn can be sorbed on the apatite when sulfide is present in minor amounts (<20 mg/L). This is less expected under stoichiometric excess of sulfide, as in the present study. However, unlike this study, the metal sulfide precipitation was studied with chemically produced sulfide (Villa-Gomez et al. 2012b), and the DOM present in the biogenic sulfide (Table 6.2) could have played a role in the decrease of the supersaturation.

The EXAFS analysis also evidenced that there are differences in the local structure of the precipitates depending on the pH (Figure 6.5). A more crystallized structure is inferred at pH 5 (Figure 6.5), in agreement with the cubic shapes observed in the SEM images (Figure 6.3b) that suggest sphalerite type structures (http://webmineral.com/data/Sphalerite.shtml) as compared to the other pH values. A decrease in supersaturation results in the formation of larger and more crystalline ZnS precipitates (Bijmans et al. 2009; Mokone et al. 2010; Sampaio et al. 2010), which could be favored at pH 5, as the threshold in ZnS solubility is reported to be around this pH value (Peters et al. 1984), and by the DOM present in the biogenic sulfide (Table 6.2).

6.4.4 Implications for metal recovery in bioreactors

Metal-recovery in bioreactors ultimately depends on the ability to separate the particles from the process wastewater (Wilson 2005). The small particle size of the metal sulfides (Figure 6.2-(Lewis 2010) is an important bottleneck towards the biotechnological application of sulfate reducing bioreactors for metal recovery in AMD and metallurgical waste streams. Such particle sizes are difficult to settle using standard clarifiers for solid state separation (Federation 2005). Crystal growth can be favoured when the supersaturation is decreased (Mersmann 1999) or agglomeration is promoted (Al-Tarazi et al. 2004; Esposito et al. 2006; Sampaio et al. 2009). In this study, those processes were found to be mediated by the DOM contained in the biogenic sulfide (Table 6.2). Despite the rather small size of the crystals formed under the conditions evaluated in this study (Figure 6.3b), clearly, further research on the characteristics of the DOM produced in sulfate reducing reactors promoting larger metal sulfide precipitates should be done. This approach has been previously used to elucidate fluorescent characteristics and metal binding properties of individual molecular weight fractions of DOM in landfill leachate (2012). These binding characteristics could, nevertheless, modify the supersaturation, thus mediating crystallization and agglomeration.

In the literature, other approaches to reduce the supersaturation have been also explored. The use of membranes to spread the sulfide dosing points not only reduced the local supersaturation, and consequently increased the size of the metal sulfide precipitates (from 10 to 22 μm), but also contributed to the agglomeration process yielding particles in the 100-200 μm range in a study of Zn and Ni precipitation with Na_2S in a continuously stirred tank reactor (Sampaio et al. 2010). One none explored approach in bioreactors is the use of a "seeding material" to promote crystal growth and thus to increase the size of the precipitates (Mersmann 1999) as used in chemical precipitators for the precipitation of heavy metals on the sand surface using sulfide or carbonate (Zhou et al. 1999; van Hille et al. 2005). This approach can be extrapolated for its application in bioreactors by using the sulfide producing biomass as seeding material to enhance crystal growth. A hint on the assessment of this approach can be observed in a previous study by Villa-Gomez et al. (Villa-Gomez et al. 2011), where the metal (Zn, Cd, Cu, Pb) sulfides accumulated in the polyethylene beads coated with SRB biofilm till the point where these beads became so heavy that they lost their floating characteristic in an inverse fluidized bed reactor.

6.5 Conclusions

- Cu and Zn sulfide precipitation characteristics are affected by the pH-dependant speciation of compounds present in the biogenic sulfide effluent such as phosphate and DOM.

- Evidence is presented that suggests that the presence of DOM might decrease the supersaturation due to binding or complexation of metal ions affecting nucleation, growth and agglomeration in metal sulfide precipitation.

- The increase of the pH displayed faster settling rate and capillary suction time as well as higher concentration of settleable solids that were linked to the DOM and macronutrients affecting nucleation, crystallization and agglomeration.

- The XANES analysis confirmed the precipitation of Zn-S as sphalerite and Cu-S as covellite in all the samples but also revealed the presence of minor amounts of Zn-sorbed on hydroxyapatite. The EXAFS analysis showed that at pH 5 a more crystallized ZnS structure is formed as in agreement with the SEM images where the cubic forms confirm sphalerite type structures, while amorphous CuS structures were observed regardless the pH.

References

Al-Tarazi M, Heesink ABM, Azzam MOJ, Yahya SA, Versteeg GF (2004) Crystallization kinetics of ZnS precipitation; an experimental study using the mixed-suspension-mixed-product-removal (MSMPR) method. Crystal Research and Technology 39 (8):675-685. doi:10.1002/crat.200310238

Al-Tarazi M, Heesink ABM, Versteeg GF, Azzam MOJ, Azzam K (2005) Precipitation of CuS and ZnS in a bubble column reactor. AIChE Journal 51 (1):235-246

APHA APHA (2005) Standard methods for examination of water and wastewater. 20 edn., Washington D.C.

Bala Subramanian S, Yan S, Tyagi RD, Surampalli RY (2010) Extracellular polymeric substances (EPS) producing bacterial strains of municipal wastewater sludge: Isolation, molecular identification, EPS characterization and performance for sludge settling and dewatering. Water Research 44 (7):2253-2266. doi:http://dx.doi.org/10.1016/j.watres.2009.12.046

Bhatia D, Bourven I, Simon S, Bordas F, van Hullebusch ED, Rossano S, Lens PNL, Guibaud G (2013) Fluorescence detection to determine proteins and humic-like substances fingerprints of exopolymeric substances (EPS) from biological sludges performed by size exclusion chromatography (SEC). Bioresource Technology 131 (0):159-165. doi:http://dx.doi.org/10.1016/j.biortech.2012.12.078

Bijmans MFM, van Helvoort P-J, Buisman CJN, Lens PNL (2009) Effect of the sulfide concentration on zinc bio-precipitation in a single stage sulfidogenic bioreactor at pH 5.5. Separation and Purification Technology 69 (3):243-248

Borsboom M, Bras W, Cerjak I, Detollenaere D, Glastra van Loon D, Goedtkindt P, Konijnenburg M, Lassing P, Levine YK, Munneke B, Oversluizen M, van Tol R, Vlieg E (1998) The Dutch-Belgian beamline at the ESRF. Journal of Synchrotron Radiation 5 (3):518-520. doi:doi:10.1107/S0909049597013484

Brooks CS (1991) Metal recovery from industrial wastes. Lewis Publishers Inc., Chelsea, MI, USA

Brown GEJ, Calas G (2012) Mineral-Aqueous solution interfaces and their impact on the Environment. Geochemical Perspectives Geochemical Perspectives 1 (4-5):483-742

Cord-Ruwisch R (1985) A quick method for the determination of dissolved and precipitated sulfides in cultures of sulfate-reducing bacteria. Journal of Microbiological Methods 4 (1):33-36

Chen W, Westerhoff P, Leenheer JA, Booksh K (2003) Fluorescence Excitation−Emission Matrix Regional Integration to Quantify Spectra for Dissolved Organic Matter. Environmental Science & Technology 37 (24):5701-5710. doi:10.1021/es034354c

Chen Y, Yang H, Gu G (2001) Effect of acid and surfactant treatment on activated sludge dewatering and settling. Water Research 35 (11):2615-2620. doi:http://dx.doi.org/10.1016/S0043-1354(00)00565-0

Djedidi Z, Khaled JB, Cheikh RB, Blais J-F, Mercier G, Tyagi RD (2009) Comparative study of dewatering characteristics of metal precipitates generated during treatment of monometallic solutions. Hydrometallurgy 95 (1-2):61-69. doi:10.1016/j.hydromet.2008.04.014

Ehrlich S, Butler I, Halicz L, Rickard D, Oldroyd A, Matthews A (2004) Experimental study of the copper isotope fractionation between aqueous Cu(II) and covellite, CuS. Chemical Geology 209 (3–4):259-269. doi:http://dx.doi.org/10.1016/j.chemgeo.2004.06.010

Esposito G, Veeken A, Weijma J, Lens PNL (2006) Use of biogenic sulfide for ZnS precipitation. Separation and Purification Technology 51 (1):31-39

Farges F (2001) Crystal chemistry of iron in natural grandidierites: an X-ray absorption fine-structure spectroscopy study Physics and Chemistry of Minerals 28:619-629

Farges F, Brown GE, Petit P-E, Munoz M (2001) Transition elements in water-bearing silicate glasses/melts. part I. a high-resolution and anharmonic analysis of Ni coordination environments in crystals, glasses, and melts. Geochimica et Cosmochimica Acta 65 (10):1665-1678

Farges F, Etcheverry M-P, Scheidegger A, Grolimund D (2006) Speciation and weathering of copper in "copper red ruby" medieval flashed glasses from the Tours cathedral (XIII century). Applied Geochemistry 21 (10):1715-1731. doi:http://dx.doi.org/10.1016/j.apgeochem.2006.07.008

Gadd GM (2009) Biosorption: critical review of scientific rationale, environmental importance and significance for pollution treatment. Journal of Chemical Technology & Biotechnology 84 (1):13-28. doi:10.1002/jctb.1999

Gallegos-Garcia M, Celis LB, Rangel-Méndez R, Razo-Flores E (2009) Precipitation and recovery of metal sulfides from metal containing acidic wastewater in a sulfidogenic down-flow fluidized bed reactor. Biotechnology and Bioengineering 102 (1):91-99

Gazea B, Adam K, Kontopoulos A (1996) A review of passive systems for the treatment of acid mine drainage. Minerals Engineering 9 (1):23-42. doi:10.1016/0892-6875(95)00129-8

Goh SW, Buckley AN, Lamb RN (2006) Copper(II) sulfide? Minerals Engineering 19 (2):204-208. doi:http://dx.doi.org/10.1016/j.mineng.2005.09.003

Hammack RW, Edenborn HM, Dvorak DH (1994) Treatment of water from an open-pit copper mine using biogenic sulfide and limestone: A feasibility study. Water Research 28 (11):2321-2329

Johnson DB, Hallberg KB (2005) Biogeochemistry of the compost bioreactor components of a composite acid mine drainage passive remediation system. Science of The Total Environment 338 (1–2):81-93. doi:10.1016/j.scitotenv.2004.09.008

Johnson KJ, Szymanowski JES, Borrok D, Huynh TQ, Fein JB (2007) Proton and metal adsorption onto bacterial consortia: Stability constants for metal–bacterial surface complexes. Chemical Geology 239 (1–2):13-26. doi:10.1016/j.chemgeo.2006.12.002

Kaksonen AH, Puhakka JA (2007) Sulfate Reduction Based Bioprocesses for the Treatment of Acid Mine Drainage and the Recovery of Metals. Engineering in Life Sciences 7 (6):541-564

Kaksonen AH, Riekkola-Vanhanen ML, Puhakka JA (2003) Optimization of metal sulphide precipitation in fluidized-bed treatment of acidic wastewater. Water Research 37 (2):255-266

Karbanee N, van Hille RP, Lewis AE (2008) Controlled Nickel Sulfide Precipitation Using Gaseous Hydrogen Sulfide. Industrial & Engineering Chemistry Research 47 (5):1596-1602. doi:10.1021/ie0711224

Lai CL, Lin SH (2004) Treatment of chemical mechanical polishing wastewater by electrocoagulation: system performances and sludge settling characteristics. Chemosphere 54 (3):235-242. doi:http://dx.doi.org/10.1016/j.chemosphere.2003.08.014

Lenz M, Hullebusch EDv, Farges F, Nikitenko S, Corvini PFX, Lens PNL (2011) Combined Speciation Analysis by X-ray Absorption Near-Edge Structure Spectroscopy, Ion Chromatography, and Solid-Phase Microextraction Gas Chromatography−Mass Spectrometry To Evaluate Biotreatment of Concentrated Selenium Wastewaters. Environmental Science and Technology 45 (3):1067-1073

Lewis A, Swartbooi A (2006) Factors Affecting Metal Removal in Mixed Sulfide Precipitation. Chemical Engineering & Technology 29 (2):277-280

Lewis AE (2010) Review of metal sulphide precipitation. Hydrometallurgy 104 (2):222-234

Luther GW, Theberge SM, Rozan TF, Rickard D, Rowlands CC, Oldroyd A (2002) Aqueous copper sulfide clusters as intermediates during copper sulfide formation, vol 36. vol 3.

Maeng SK, Sharma SK, Abel CDT, Magic-Knezev A, Amy GL (2011) Role of biodegradation in the removal of pharmaceutically active compounds with different bulk organic matter characteristics through managed aquifer recharge: Batch and column studies. Water Research 45 (16):4722-4736. doi:10.1016/j.watres.2011.05.043

Mokone TP, van Hille RP, Lewis AE (2010) Effect of solution chemistry on particle characteristics during metal sulfide precipitation. Journal of Colloid and Interface Science 351 (1):10-18. doi:10.1016/j.jcis.2010.06.027

Moreau JW, WEBB, I. R, BANFIELD, F. J (2004) Ultrastructure, aggregation-state, and crystal growth of biogenic nanocrystalline sphalerite and wurtzite, vol 89. vol 7. Mineralogical Society of America, Washington, DC, ETATS-UNIS

Moreau JW, Weber PK, Martin MC, Gilbert B, Hutcheon ID, Banfield JF (2007) Extracellular Proteins Limit the Dispersal of Biogenic Nanoparticles. Science 316 (5831):1600-1603. doi:10.1126/science.1141064

Muñoz M, Argoul P, Farges F (2003) Continuous Cauchy wavelet transform analyses of EXAFS spectra: A qualitative approach, vol 88. vol 4. Mineralogical Society of America, Washington, DC, ETATS-UNIS

Neculita C-M, Zagury GJ, Bussiere B (2007) Passive Treatment of Acid Mine Drainage in Bioreactors using Sulfate-Reducing Bacteria: Critical Review and Research Needs. J Environ Qual 36 (1):1-16. doi:10.2134/jeq2006.0066

Ni B-J, Fang F, Xie W-M, Sun M, Sheng G-P, Li W-H, Yu H-Q (2009) Characterization of extracellular polymeric substances produced by mixed microorganisms in activated sludge with gel-permeating chromatography, excitation–emission matrix fluorescence spectroscopy measurement and kinetic modeling. Water Research 43 (5):1350-1358

Parsons JG, Hejazi M, Tiemann KJ, Henning J, Gardea-Torresdey JL (2002) An XAS study of the binding of copper(II), zinc(II), chromium(III) and chromium(VI) to hops biomass. Microchemical Journal 71 (2):211-219. doi:10.1016/s0026-265x(02)00013-9

Pattrick RAD, Mosselmans JFW, Charnock JM, England KER, Helz G, Garner R, Vaughan DJ (1997) The structure of amorphous copper sulfide precipitates: An X- ray absorption study Geochimica et Cosmohimica Acta 61 (10):2023-2036

Peters RW, Chang T-K, Ku Y (1984) Heavy metal crystallization kinetics in an MSMPR crystallizer employing sulfide precipitation. Journal Name: AIChE Symp Ser; (United States); Journal Volume: 80:240:Medium: X; Size: Pages: 55-75

Piccolo A (2002) The supramolecular structure of humic substances: A novel understanding of humus chemistry and implications in soil science. In: Advances in Agronomy, vol Volume 75. Academic Press, pp 57-134

Prange A, Modrow H (2002) X-ray absorption spectroscopy and its application in biological, agricultural and environmental research. Reviews in Environmental Science and Biotechnology 1 (4):259-276. doi:10.1023/a:1023281303220

Raynaud M, Vaxelaire J, Olivier J, Dieudé-Fauvel E, Baudez J-C (2012) Compression dewatering of municipal activated sludge: Effects of salt and pH. Water Research 46 (14):4448-4456. doi:http://dx.doi.org/10.1016/j.watres.2012.05.047

Rickard DT (1972) Covellite formation in low temperature aqueous solutions. Mineral Deposita 7 (2):180-188. doi:10.1007/bf00207153

Sahinkaya E, Gungor M, Bayrakdar A, Yucesoy Z, Uyanik S (2009) Separate recovery of copper and zinc from acid mine drainage using biogenic sulfide. Journal of Hazardous Materials 171 (1-3):901-906

Sampaio RMM, Timmers RA, Kocks N, André V, Duarte MT, van Hullebusch ED, Farges F, Lens PNL (2010) Zn–Ni sulfide selective precipitation: The role of supersaturation. Separation and Purification Technology 74 (1):108-118. doi:10.1016/j.seppur.2010.05.013

Sampaio RMM, Timmers RA, Xu Y, Keesman KJ, Lens PNL (2009) Selective precipitation of Cu from Zn in a pS controlled continuously stirred tank reactor. Journal of Hazardous Materials 165 (1-3):256-265

Shea D, Helz GR (1989) Solubility product constants of covellite and a poorly crystalline copper sulfide precipitate at 298 K. Geochimica et Cosmochimica Acta 53 (2):229-236. doi:http://dx.doi.org/10.1016/0016-7037(89)90375-X

Sheng G-P, Yu H-Q (2006) Characterization of extracellular polymeric substances of aerobic and anaerobic sludge using three-dimensional excitation and emission matrix fluorescence spectroscopy. Water Research 40 (6):1233-1239. doi:10.1016/j.watres.2006.01.023

Sobeck DC, Higgins MJ (2002) Examination of three theories for mechanisms of cation-induced bioflocculation. Water Research 36 (3):527-538. doi:http://dx.doi.org/10.1016/S0043-1354(01)00254-8

Veeken AHM, Akoto L, Hulshoff Pol LW, Weijma J (2003) Control of the sulfide (S2-) concentration for optimal zinc removal by sulfide precipitation in a continuously stirred tank reactor. Water Research 37 (15):3709-3717

Villa-Gomez D, Ababneh H, Papirio S, Rousseau DPL, Lens PNL (2011) Effect of sulfide concentration on the location of the metal precipitates in inversed fluidized bed reactors. Journal of Hazardous Materials 192 (1):200-207. doi:10.1016/j.jhazmat.2011.05.002

Villa-Gomez DK, Papirio S, van Hullebusch ED, Farges F, Nikitenko S, Kramer H, Lens PNL (2012) Influence of sulfide concentration and macronutrients on the characteristics of metal precipitates relevant to metal recovery in bioreactors. Bioresource Technology 110 (0):26-34. doi:10.1016/j.biortech.2012.01.041

Wilson TE (2005) Introduction and overview water environmental federation clarifier design, 2nd ed. Manual of Practice No. FD-8. Proceedings of the Water Environment Federation 2005 (11):4412-4416. doi:10.2175/193864705783866810

Winterer M (1997) XAFS - A Data Analysis Program for Materials Science. J Phys IV France 7 (C2):C2-243-C242-244

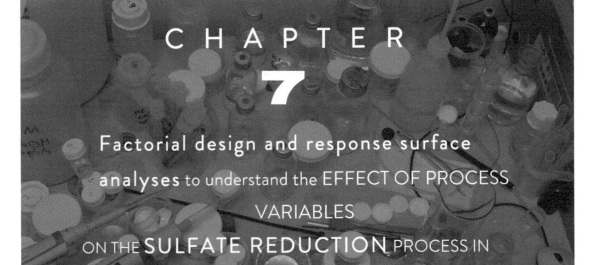

CHAPTER
7

Factorial design and response surface analyses to understand the EFFECT OF PROCESS VARIABLES ON THE **SULFATE REDUCTION** PROCESS IN **THE IFB REACTOR**

Abstract

The individual and combined effect of the pH, chemical oxygen demand (COD) and SO_4^{2-} concentration, metal to sulfide (M/S^{2-}) ratio and hydraulic retention time (HRT) on the biological sulfate reduction (SR) process was evaluated in an inverse fluidized bed reactor by factorial design analysis (FDA) and response surface analysis (RSA). The regression-based model of the FDA described the experimental results well and revealed that the most significant variable affecting the process was the pH. The combined effect of the pH and HRT was barely observable, while the pH and COD concentration positive effect (up to 7 and 3 gCOD/L, respectively) enhanced the SR process. Contrary, the individual COD concentration effect only enhanced the COD removal efficiency, suggesting changes on the microbial pathway. The RSA showed that the M/S^{2-} ratio determined whether the inhibition mechanism to the SR process was due to the presence of free metals or precipitated metal sulfides.

This Chapter was submitted for publication as:

Villa-Gomez D. K., Mushi S., Maestro R., Pakshirajan K. and Lens P.N.L. (2013) Effect of process variables on the sulfate reduction process in bioreactors treating metal-containing wastewaters: factorial design and response surface analyses. Submitted to *Journal of Environmnetal Technology.*

7.1 Introduction

Waste and process water from the mining and metallurgical industry typically contain
high concentrations of dissolved metal-ions and sulfuric acid (Johnson and Hallberg
2005). Although these acidic waste streams are a major environmental problem, they
contain valuable metals that can be recovered by processes such as metal sulfide
precipitation (Huisman et al. 2006; Tabak et al. 2003). Sulfate, which is commonly
found in these wastewaters, can be easily transformed to sulfide by sulfate reducing
bacteria (SRB) (Kaksonen and Puhakka 2007).

Simultaneous biological sulfate reduction and metal sulfide precipitation can be
achieved in a single stage process that aids to reduce the number of process units.
However, such a process requires careful monitoring of the sulfate reduction process to
avoid microbial inhibition due to either metals toxicity or low pH conditions in the
bioreactor. It is also necessary to maintain an acceptable process performance and to
avoid failure in case of shock loadings. In addition, metal sulfide precipitates are
difficult to recover from the bioreactor because the metal precipitation occurs nearby or
within the biomass, which thus necessitates an additional unit operation to separate the
biomass from the precipitated metals. Consequently, biomass may be lost in the process
leading to reduced system performance. These drawbacks could be overcome by
employing an inverse fluidized bed (IFB) reactor that enables sulfide production by
SRB and metal recovery in the same unit due to its floating biomass bed (Gallegos-
Garcia et al. 2009; Villa-Gomez et al. 2011).

Over the last few decades, different sulfate reducing bioreactor configurations have
been used for the treatment of metal-containing wastewaters in a single stage (Kaksonen
and Puhakka 2007; Neculita et al. 2007; Bijmans et al. 2011; Papirio et al. 2012).
However, the operational conditions such as pH, hydraulic retention time (HRT),
electron donor and metals (type and concentration) applied in these studies vary too
significantly that comparison of process performance across these studies is
complicated. The individual effect of different operational parameters such as the pH,
HRT, electron donor concentration and metal to sulfide molar (M/S^{2-}) ratio in sulfate
reducing bioreactors has been reported (Kaksonen and Puhakka 2007). In the case of
pH, simultaneous biological sulfate reduction and metal precipitation is reported at the
lowest pH of 5 in the influent and effluent (Bijmans et al. 2008a; Gallegos-Garcia et al.

2009), while sulfate reduction without metal addition has been reported at pH 3.8 in a batch reactor fed with glycerol (Kimura et al. 2006) and at pH 4.0 in a sucrose fed upflow anaerobic sludge blanket (UASB) reactor (Lopes et al. 2007) as well as in a formate/hydrogen fed membrane bioreactor (Bijmans et al. 2010). The study of the effect of process parameters is particularly important from a standpoint of understanding the role of the different variables on the sulfate reduction process. However, the data obtained in these studies are based on experiments performed by manipulating the variables one-at-a-time and neglect the effect of multi-variables on the sulfate reduction process.

Factorial design analysis is a useful tool to interpret the effect of multiple parameters, especially when interactions between the process variables are involved (Montgomery 2004). This tool has been applied to study certain biological processes and has not only yielded a better interpretation of the results, but also a reduction of the number of experiments to be performed. For instance, White and Gadd (1996) studied the metal removal and alkalization by SRB of a typical acid leachate containing metals from contaminated soils. Key variables affecting the process were found to be the type of substrate, the chemical oxygen demand (COD)/SO_4^{2-} ratio and the interaction between these variables along with the dilution rate.

The objective of this study was to understand the effect of significant operational parameters on the sulfate reduction process in an IFB reactor for the treatment of metal containing wastewater. The IFB reactor was operated at different influent COD (concomitantly SO_4^{2-}) concentration and HRT, varying thus the organic loading rate (OLR), as well as different M/S^{2-} ratio and pH for the identification of the main and interaction effects and their significances on the sulfate reduction process, as given by a factorial design and response surface analysis of the results obtained.

7.2 Materials and methods

7.2.1 Bioreactor set up and experimental design

All bioreactor experiments were conducted employing two simultaneously operated IFB reactors as described in detail by Villa-Gomez et al. (2012a). All the reactor runs were performed at 25 ($\pm$3) °C and at a fixed recirculation liquid down flow rate. The COD/SO_4^{2-} ratio for runs 1-10 was 0.67, while for runs 11-16 the COD/SO_4^{2-} ratio was 1. A SRB biofilm was established, yielding incomplete oxidation of lactate producing acetate and sulfide as the end products (Villa-Gomez et al. 2012a) and no further biomass was added during this study.

The individual and combined effect of the COD_{in} concentration, pH, M/S^{2-} ratio and HRT on the effluent sulfide concentration, COD and sulfate removal efficiency were evaluated in a total of 16 reactor runs (Table 7.1). These three responses were chosen as they indicate the sulfate reduction efficiency. The levels of the COD_{in} concentration, pH (inside the reactor), M/S^{2-} ratio and HRT were varied in the range 1-3 g/L, 4-7, 0-2.6 and 4.5-24 h, respectively. The sulfate concentration varied proportionally with the COD_{in} concentration to keep the COD/SO_4^{2-} ratio constant. The effect of the OLR was not evaluated separately as it depends on two variables (HRT and COD_{in}) that were already included in the parameter study. The parameter M/S^{2-} ratio, shown in Table 7.1, was calculated from the effluent sulfide concentration in each run at steady state conditions.

Table 7.1 Operational conditions in the different experimental reactor runs.

Run	Operation time (d)	COD_{in} (gCOD/L)	pH	HRT (h)	Sulfide (mg/L)	M/S^{2-} ratio (mol/mol)
1	52	2.0	7.0	24.0	413.4	0.0
2	40	2.0	6.5	24.0	366.6	0.0
3	26	2.0	6.0	24.0	178.6	0.0
4	23	2.0	5.0	24.0	170.0	0.0
5	26	2.0	4.0	24.0	42.7	0.0
6	28	2.0	7.0	24.0	176.0	0.8[b]
7	38	1.0	5.0	24.0	61.0	0.8[b]
8	25	1.0	4.0	24.0	23.0	0.8[b]
9	30	1.0	7.0	24.0	253.3	0.0
10	20	1.0	4.0	24.0	25.1[a]	0.0
11	58	2.0	7.0	24.0	287.2[a]	0.5[c]
12	47	2.0	7.0	9.0	301.9	0.5[c]
13	34	2.0	7.0	4.5	54.9	2.6[c]
14	58	3.0	7.0	24.0	540.0	0.3[c]
15	47	3.0	7.0	9.0	379.0	0.4[c]
16	34	3.0	7.0	4.5	146.5	1.0[c]

[a]Calculated from the sulfate removal efficiency results, M/S^{2-} ratio: [b]$\Sigma(Zn^{2+},Cu^{2+})/[S^{2-}]$ and [c]$\Sigma(Zn^{2+},Cd^{2+}, Pb^{2+}, Cu^{2+})/[S^{2-}]$.

The operation time of the reactors depended on the time needed to reach pseudo-steady state conditions of the sulfide concentration values for at least $5 \times$ HRT at the conditions of each run. Thus, the duration of each experimental run was in the range of 20-51 days (Table 7.1). The experimental runs 1-11 were performed by varying the levels of pH and COD_{in} concentration one-at-a-time and by keeping the HRT constant, whereas the results from Villa-Gomez et al. (2012a) were included to study the effect of HRT (runs 12-16).

7.2.2 Synthetic wastewater

The synthetic wastewater fed to the IFB reactors contained (mg/L): KH_2PO_4 500, NH_4Cl 200, $CaCl_2·2H_2O$ 2500, $FeSO_4·7H_2O$ 50 and $MgSO_4·7H_2O$ 2500 and micronutrients as described by Zehnder et al. (1980). Sulfate was supplemented as Na_2SO_4, whereas $NaC_3H_5O_3$ and $C_3H_6O_3$ were used for lactate. Zn and Cu were added to the influent in the runs 6-8 and Zn, Cd, Pb and Cu were added to the influent in runs 11-16 (Table 7.1), for which the respective chloride salts were used. When metals were added in the influent, the yeast extract and the micronutrients were eliminated from the synthetic medium as these can interfere with the metal sulfide precipitation. All reagents were of analytical grade.

7.2.3 pH control and monitoring

The pH inside the reactor was adjusted to the desired pH by either manual addition of NaOH or HCl to the influent or by an automated operation using a data acquisition card (DAC) (NI cDAQ-9174, National Instruments, The Netherlands) and LabView® software (version 2009). The LabView® software contained a PID controller (PID and Fuzzy Logic Toolkit, National Instruments, The Netherlands). A pH electrode (Prosense, The Netherlands) submerged in the IFB bioreactor mixed liquor and connected to the DAC was used for monitoring the reactor pH.

7.2.4 Analytical methods

COD was determined by the close reflux method (APHA 2005). Sulfate and metal determination followed the procedures as described by Villa-Gomez et al. (2011). Sulfide was determined spectrophotometrically by the colorimetric method described by Cord-Ruwisch (1985) using a UV-visible spectrophotometer (Perkin Elmer Lambda 20, USA). Acetate was measured as described by Villa-Gomez et al. (2012b).

7.2.5 Statistical analysis of the results

For the interpretation of the role of the different process variables investigated in this study, the results obtained from the experimental reactor runs were submitted to statistically valid full factorial design and response surface analyses. The results were thus analyzed in the form of analysis of variance (ANOVA), student t test and response surfaces (Montgomery 2004). For this purpose, the values of the input variables were coded by the statistical software Minitab® (Ver 14.1, PA, USA) and used as such for the analyses. Thus, -1 and +1 represented the lowest and highest levels, respectively, in the coded units of the input variables. For estimating the individual and combined effect of the variables (COD_{in} concentration, pH, HRT and M/S^{2-} ratio) and their significance, a statistical regression based model was developed using the Minitab® software. The estimated effects were relative measures of the sign and magnitude of impact of the process variables as described in the results section. The effect due to a variable was defined as either enhancement (positive) or reduction (negative) in the responses.

7.3 Results

7.3.1 Effect of pH on COD and sulfate removal efficiency and sulfide concentration

A linear correlation between the pH and the COD removal as well as between the pH and sulfate removal efficiency was observed by varying the pH from 5 to 7 in all the experimental runs (Figure 7.1a). At pH 4, the COD removal efficiency was 25, 50 and 90%, while the sulfate removal efficiencies remained similar to those achieved at pH 5 (Figure 7.1a). The difference in COD and sulfate removal efficiency (Figure 7.1a) as well as in sulfide concentration (Figure 7.1b) obtained at pH 7 is due to the differences in the other operational parameters (COD_{in} and metals concentration), which is not observable in this two dimension figure. The sulfide concentration decreased linearly with the pH decrease regardless of the metals addition (Figure 7.1b). On the contrary, the sulfide concentration was slightly higher with the metal addition in the system at pH 7 (Figure 7.1b).

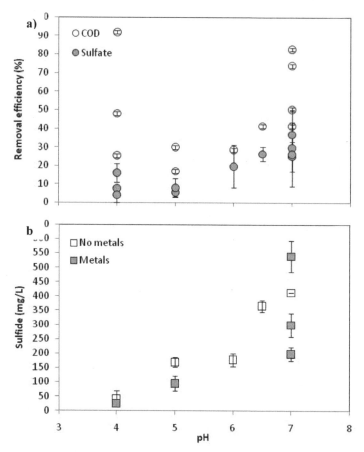

Figure 7.1 Influence of pH on a) COD and sulfate removal efficiency and b) sulfide concentration in experimental runs with and without metals during reactor operation at an HRT of 24 h (vertical bars indicate ± standard deviation).

7.3.2 Effect of HRT on the sulfide concentration produced

The comparison of the runs carried out at different HRTs (runs 11-16) underlined the increase in sulfate reducing activity when the reactors were submitted to longer HRTs (Figure 7.2). When the HRT is decreased from 24 to 9 h, the sulfide concentration slightly decreased at a COD_{in} concentration of 3 gCOD/L, while at 2 gCOD/L, the sulfide concentration is even higher at an HRT of 9 h compared to that obtained at an HRT of 24 h. The sulfide concentration greatly decreased when the HRT is decreased from 9 to 4.5 h at both the COD_{in} concentrations.

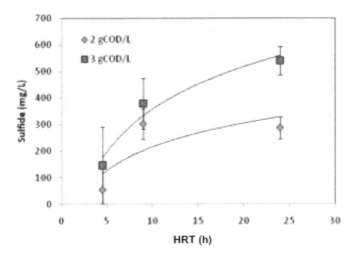

Figure 7.2 Sulfide production at different HRTs with a COD_{in} concentration of 2 and 3 gCOD/L in the presence of metals (runs 11-16). Continuous lines indicate the fitted logarithmic curve for both COD_{in} concentrations (vertical bars indicate ± standard deviation).

7.3.3 Metal removal efficiency

Table 7.2 shows the metal removal efficiencies obtained in the experimental runs 6-8, while the metal removal efficiencies for runs 11-16 are found in a previous study (Villa-Gomez et al. 2012a). In all the experimental runs with metal addition, precipitation occurred due to metal sulfide formation when the M/S^{2-} ratio was above 1, which was confirmed by speciation analysis (data not shown) using the Visual MINTEQ version 3.0 (US EPA, 1999). Thus, the variation in the metal removal efficiency can be attributed to a decrease in particle size of the metal precipitates that escaped with the effluent. At HRTs of 4.5 h in runs 13 and 16, the M/S^{2-} ratio was below 1, and therefore, the decrease in the metal removal efficiencies are attributed to the decrease in metal sulfide precipitation (Villa-Gomez et al. 2012a).

Table 7.2 Zn and Cu removal efficiencies in IFB reactor runs 6 to 8.

Runs	Zn removal efficiency (%)	Cu removal efficiency (%)
6	94.5 ± 6.4	92.7 ± 7.5
7	97.2 ± 0.75	97.9 ± 0.9
8	81.4 ± 13.8	94.4 ± 3.1

7.3.4 Statistical analysis of the results

7.3.4.1 Factorial design analysis

Statistical analysis of the results in the form of analysis of variance (ANOVA) and Student t test was performed for interpretation of the main and interactive (2 and 3-way) effects of the operational parameters on the sulfate reduction process (Table 7.3). In general, a high Fisher's (F) value with a low probability P indicates a high significance of the corresponding regression model term (Montgomery 2004). Thus, the very low P values of the main and 2-way interaction effects for sulfide concentration (0.072 and 0.058, respectively) reveal a very high significance of this response over the other two (COD and sulfate removal efficiency). After the sulfide values, the 2-way and 3-way interactions for the COD removal efficiency were more significant with greater F-values (13.01 and 11.89, respectively) as compared to the main effects (7.28).

Table 7.3 ANOVA of the sulfate reduction process showing the significance of the main and interactive effects between the process variables.

	F	P
COD removal efficiency		
Main effects	7,28	0,27
2-way interactions	13,01	0,205
3-way interactions	11,89	0,18
Sulfate removal efficiency		
Main effects	0,28	0,868
2-way interactions	0,75	0,687
3-way interactions	0,5	0,609
Sulfide concentration		
Main effects	107,04	0,072
2-way interactions	169,49	0,058
3-way interactions	0,25	0,705

In order to further ascertain which of the main and interaction effect between the process variables played significant roles in the process, a student t test was performed on all the three responses (COD and sulfate removal efficiency and sulfide concentration). Table 7.4 gives the calculated individual (main) and interactive effects between the variables for these three responses, which are reflected in the sign and magnitude of the impact of the process variable. For all the three responses, the pH had the most significant effect (P<0.1) and its effect was positive in comparison with the other variable effects. The individual effects of both the COD_{in} and M/S^{2-} ratio were also significant (P<0.1), but mainly for the sulfide concentration (Table 7.4), whereas the significance of the individual effect for the HRT was low (P>0.3), particularly for the COD removal efficiency (P= 0.9).

The individual effect of the HRT displayed negative values for the sulfide concentration and the COD removal efficiencies, while for the sulfate removal efficiency, its effect was positive. Whereas an increase in pH showed a positive effect on all the three responses, an increase in the COD_{in} concentration and the M/S^{2-} ratio was inhibitory to the sulfate reduction (sulfide concentration and sulfate removal efficiency) as indicated by their strong negative values. Conversely, the COD_{in} concentration displayed a positive main effect on the COD removal efficiency.

Table 7.4 Student 't' test to determine the significant individual and interaction effects between the variables on the sulfide concentration and the COD and sulfate removal efficiency.

Variable	Sulfide (mg/L)			COD removal (%)			Sulfate removal (%)		
	Effect	T	Significance level (P)	Effect	T	Significance level (P)	Effect	T	Significance level (P)
Main effects									
pH	362.7	5.6	0.005	47.26	2.22	0.091	27.148	3.78	0.019
HRT	-113.5	-1.2	0.307	-3.9	-0.1	0.909	3.043	0.28	0.791
COD_{in}	-496.7	-2.3	0.084	2.79	0.04	0.971	-11.352	-0.5	0.661
M/S ratio	-451.3	-3.9	0.018	-16.47	-0.4	0.69	-17.482	-1.4	0.247
Interactive effects									
pH+COD_{in}	152.7	0.65	0.549	125.26	1.63	0.179	24.148	0.93	0.403
HRT+COD_{in}	-146.1	-0.4	0.729	163.06	1.26	0.276	34.993	0.8	0.466
HRT+M/S ratio	-145.3	-1.3	0.266	-49.94	-1.4	0.249	14.917	1.2	0.297
COD_{in}+M/S ratio	-759.3	-1.4	0.229	188.19	1.07	0.346	5.924	0.1	0.925
HRT+COD_{in}+M/S ratio	-377.7	-0.7	0.502	209.3	1.24	0.283	17.849	0.31	0.769

The significance levels of the interactive effects ranged between 0.229 and 0.925. Among the interactive effects, the pH and COD_{in} concentration showed a positive effect on the sulfide concentration. For the COD removal efficiency, all the interactive effects that involved the COD_{in} concentration as one of the parameters were positive. The interactive effects for the sulfate removal efficiency were all positive, in contrast to that of the sulfide concentration.

Table 7.5 presents the estimated coefficients of individual and interaction effects between the variables on sulfide production (Y_1), COD and sulfate removal efficiency (Y_2 and Y_3, respectively), where (X_1), (X_2), (X_3) and (X_4) represent the variables COD_{in}, pH, HRT and M/S^{2-} ratio. These estimated coefficients could thus be used to represent the regression model equation for each of the responses (Y_1, Y_2, Y_3) as shown below:

$$Y = b_0 + b_1X_1 + b_2X_2 + b_3X_3 + b_4X_4 + b_5X_1X_2 + b_6X_1X_3 + b_7X_1X_4 + b_8X_2X_4 + b_9X_3X_4 + b_{10}X_1X_2X_3 \dots \dots \dots \dots \dots \dots \dots \dots \dots \dots \dots \dots \dots \dots \dots \dots \quad (1)$$

Table 7.5 Estimated coefficients (bi) from equation 1 for the sulfide production, COD and sulfate removal efficiency.

Term		Coefficient (b_i)		
		Y= Sulfide	Y=COD removal	Y=Sulfate removal
X_o (constant)	b_0	251.4	13.44	80.6
X_1	b_1	-167.1	-15.12	-307.6
X_2	b_2	-34.7	5.97	12.9
X_3	b_3	2.3	-1.34	-4.4
X_4	b_4	-91.5	159.19	1179.9
X_1X_2	b_5	23.3	1.46	57.5
X_1X_3	b_6	-0.1	0.34	3.2
X_1X_4	b_7	-12.6	-23.61	-267.6
X_2X_4	b_8	18.9	-18.2	-104.9
X_3X_4	b_9	-2.8	-4.65	-32.1
$X_1X_2X_3$	b_{10}	1.2	2.63	12.6

These coefficients (Table 7.5) and the regression (Equation 1) for each of the responses were used to obtain the model predicted values, from which the normal probability plots of the residuals and the residuals versus fitted values plots were generated to compare the accuracy of these models in predicting the experimental data (Figure 7.3).

The normal probability plots of the sulfide concentration as well as the COD and sulfate removal efficiency display an approximately linear profile (Figure 7.3) indicating the validity of the model observation of the experimental data (Montgomery, 2004). However, a negative value for both the COD removal efficiency and the sulfide concentration as well as a positive value for the sulfate removal efficiency deviate from the linear tendency (Figure 7.3). These off-set values are also observed in the residual plots that correspond to the predicted values for run 11. This unusual observation is due to the low efficiency values obtained during this reactor run despite the favorable operational pH (Table 7.1). A probable reason for this unusual observation in run 11 could be that the previous reactor runs were performed at a low pH conditions from which the biomass was not recuperated quickly enough.

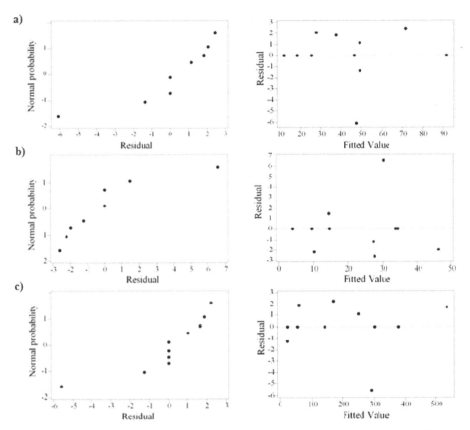

Figure 7.3 Normal probability (left) and residual (right) plots showing the difference between the observed and the model predicted values for a) COD removal efficiency (%), b) sulfate removal efficiency (%) and c) sulfide concentration (mg/L).

7.3.4.2 Response surface analysis

The regression equations obtained using the coefficients displayed in Table 7.5 for the COD_{in} concentration, HRT and M/S^{2-} ratio were used to generate matrices of data for constructing two dimension contour plots on the three responses as a function of the pH and the other variables tested (Figure 7.4). The response surfaces for the sulfide production and the sulfate removal efficiency show a decrease with decrease of both the pH and COD_{in} concentration, while no linear tendency is observed for the M/S^{2-} ratio (Figure 7.4a and 7.4c). The contour plots between pH and the COD_{in} concentration for the three responses confirm the strong interactive effect of these two parameters as also revealed by the ANOVA and the student t test results (Table 7.3). However, unlike the strong negative interaction between the two parameters on the sulfide concentration and sulfate removal efficiency, the individual effect of pH predominated over that of COD_{in} for the COD removal efficiency. For instance, with a decrease of the pH in the reactor below 5, the COD removal efficiency decreased (35%), whereas an increase above pH 5 improved the COD removal efficiency (65%) (Figure 7.4b left).

The interaction effect between the pH and HRT on the three responses is barely observable (Figure 7.4 middle), and the contours of the sulfide concentration and sulfate removal efficiency are strongly defined by the pH in the reactor. Similarly, the COD removal efficiency did not vary much following a change in these two parameters (Figure 7.4b middle).

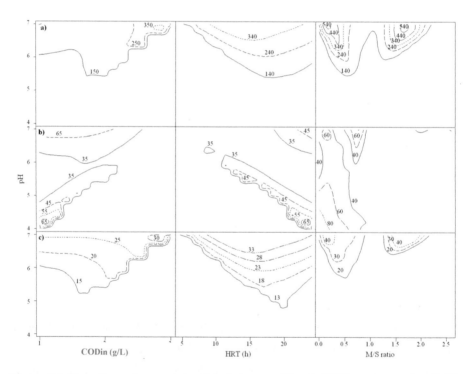

Figure 7.4 Two dimension contour plots between pH and COD_{in} concentration (left), HRT (middle) and M/S^{2-} ratio (right) for a) sulfide production, b) COD removal efficiency and c) sulfate removal efficiency. The data matrix was generated using the coefficients presented in Table 7.5.

The combined effect of the pH and M/S^{2-} ratio shows two regions where the change in the sulfide concentration and the sulfate removal efficiency vary considerably (Figure 7.4a and 7.4c right). These regions are located at the M/S^{2-} ratio below 0.5 and above 1.5. The increase in the M/S^{2-} ratio up to 1 decreases the sulfide concentration and the pH effect is less strong, while the effect of the M/S^{2-} ratio above 1 is less strong but it is more sensitive to the pH (Figure 7.4c, right). It is interesting to observe that the COD removal efficiency is highly affected by the M/S^{2-} ratio, while the pH effect is less pronounced on this response (Figure 7.4b right).

7.4 Discussion

7.4.1 Statistical interpretation on the effect of process variables

The statistical analysis (Tables 7.4, 7.5 and Figure 7.3) indicated that the regression-based model described the experimental results well (Montgomery 2004). Among the different process variables evaluated in this study (pH, COD_{in} concentration, HRT and M/S^{2-} ratio), the most significant variable affecting the sulfate reduction process was found to be the pH (Table 7.4, Figure 7.4), whereas the other variables affected mainly the sulfide concentration response. Other than the individual effects, two or three way interaction effects between the variables were only slightly significant or not significant at all (Tables 7.4 and 7.5).

The main and interactive effects of the variables on the sulfide concentration were found to be the most significant compared to the other two responses (COD and sulfate removal efficiency), revealing that the sulfide concentration in the system was the most reliable response for assessing the sulfate reduction efficiency in the IFB bioreactor. In contrast, the significance of the sulfate removal efficiency was notably low, suggesting that small amounts of the sulfate could be transformed due to other mechanisms apart from biological sulfate reduction process. These mechanisms include: sulfate precipitation with the synthetic wastewater components (Villa-Gomez et al. 2012a), biological sulfide re-oxidation to sulfate due to air intrusion into the bioreactor (Canfield and Raiswell 1999) and possibly anaerobic oxidation by photosynthetic bacteria (Bruser et al. 2000) as the reactors were not shielded from light. The low significance observed in the response of the COD removal efficiency could be attributed to the fact that microorganisms, such as fermentative bacteria and methane producing bacteria, compete with SRB for the electron donors (Omil et al. 1998; Oyekola et al. 2009).

7.4.2 Effect of the pH

This study shows that the main factor affecting the process is the pH, while the effect of the other variables is of less importance on the sulfate reducing process (Table 7.4, Figure 7.4). Several studies have reported a notable decrease of the sulfate reducing activity at low pH values in bioreactors regardless of the differences in substrate type and concentration (Kimura et al. 2006; Lopes et al. 2007) or bioreactor configuration (Bijmans et al. 2010). In addition, various operational conditions such as the metal sulfide solubility product and sulfide species (H_2S, HS^-, S^{2-}) that can potentially cause metal and sulfide toxicity are influenced by the pH (Petrucci and Moews, 1962). Nevertheless, this study also showed differences in both the individual as well as the combined effect of the pH with other process variables on the responses (Table 7.4, Figure 7.4).

The response surface analysis showed that the combined effect of the pH differed between the COD removal efficiency and the sulfate reduction (sulfide concentration and sulfate removal efficiency) (Figure 7.4b and 7.4c). The contour plots clearly show that the COD removal efficiency is affected with the decrease of pH up to 5, and above this pH value, the effect was reversed with slight differences due to its combined effect with other parameters (COD_{in} concentration, HRT or M/S^{2-} ratio) (Figure 7.4b). Microorganisms other than the SRB present in the bioreactor can contribute to the substrate consumption at pH 4 (Figure 7.4b). For instance, acidic conditions are needed to induce the activity of lactic-acid-degrading *L. buchneri* (Elferink et al. 2001). At a pH above 5.8, lactic acid consumption was negligible, whereas cells grown at pH 4.3 showed an increased conversion rate of lactic acid with very high conversion rates when grown at pH 3.8 (Elferink et al. 2001). In contrast, the sulfate removal efficiency and sulfide concentration decreased linearly with the pH in the range studied (Figure 7.4). The toxicity of HS^- to SRB and other microbes is less than that of H_2S (Mora-Naranjo et al. 2003), which predominates under the acidic conditions (pH<5) (Petrucci and Moews, 1962) tested in this study (Table 7.1).

7.4.3 Effect of the M/S^{2-} ratio

The response surface analysis indicates that the three responses are affected by the M/S^{2-} ratio (Figure 7.4a and 7.4c right). Furthermore, although the ANOVA results (Table 7.3) did not reveal a significant interaction between the pH and the M/S^{2-} ratio, the response surface analysis showed a significant combined effect between these two factors on the sulfate reduction process (Figure 7.4a and 7.4c right). The combined effect of the pH and the M/S^{2-} ratio is also due to the fact that the pH determines the solubility of the metal species and, therefore, their availability in solution (Labrenz et al. 2000; Gonzalez-Silva et al. 2009). Since the metals tested in this study (Cd, Zn, Cu, Pb) are not soluble at a pH below 4 (Petrucci and Moews, 1962), which is the lowest pH tested in this study (Table 7.1), their toxic effect is neglected under excess of sulfide. As a consequence, the effect of the metal addition depended whether the M/S^{2-} ratio was above or below the M/S^{2-} ratio of 1 (Figure 7.4a and 7.4c right). At a M/S^{2-} ratio below 1, the amount of sulfide produced is high thus ensuring metal sulfide precipitation and therefore any toxic effects by free metals can be neglected (Kaksonen et al. 2006). However, inhibition of sulfate reduction by insoluble metal sulfides can still occur (Utgikar et al. 2004; Utgikar et al. 2002), particularly at pH values below neutral (Reis et al. 1992; Moosa and Harrison 2006). In case the M/S^{2-} ratio is higher than 1, the probability of free metal toxicity in the system is high, resulting in a decrease of the sulfate removal efficiency and thus the sulfide concentration (Figure 7.4a and 7.4c right).

Unlike sulfate reduction, the COD removal efficiency highly depended on the M/S^{2-} ratio but not on the pH (Figure 7.4b right). Metals can stimulate or inhibit certain anaerobic microorganisms depending on process-related factors such as pH and redox potential (Chen et al. 2008). For example, acetogens are more resistant to heavy metal toxicity than methanogens (Zayed and Winter 2000). This is important since the outcome of the competition between SRB and other anaerobic microorganisms determines the sulfide concentration produced in the reactor. This study thus evidences that the sulfate reduction efficiency is less affected by the metal addition (Figure 7.4a and 7.4c, right) as compared with the strong effect caused by the pH change (Table 7.4, Figure 7.4a and 7.4 c, right).

7.4.4 Combined effect of the pH and COD$_{in}$ concentration

The combined effect of the pH and COD$_{in}$ concentration seems to be positive for the sulfate reduction efficiency when both parameters are set at their high levels (pH 7 and 3 gCOD/L-Figure 7.4a and 7.4b left), whereas the individual effect due to COD$_{in}$ concentration was not at all significant on the sulfate removal efficiency (Table 7.3). Due to its role as carbon/energy source, the COD concentration is known to enhance the microbial growth rate; however, in this study it was shown that its effect is only positive when the pH is also high (Figure 7.4 left). This synergistic effect is due to an excess of COD in the system, which favors the diversity of the microbial species present (Oyekola et al. 2009) that can help maintain sulfate reduction even under adverse conditions such as low pH. Similarly, Lopes et al. (2007) reported a nearly complete sulfate reduction efficiency during the acidification of sucrose at pH 6 and 5 at influent COD to SO$_4^{2-}$ ratios of 9 and 3.5, respectively.

7.4.5 Effect of the HRT

The statistical analysis shows that the HRT parameter had barely any effect on the sulfate reducing process (Table 7.3 and Figure 7.4 middle). However, the experimental data show that the sulfide concentration does not increase at HRT values above 9 h (Figure 7.2). This is because the sulfide production rate depends on the kinetics of the microbial community and the quantity of SRB biomass in the bioreactor. Therefore, operating at an HRT above this value might be unnecessary for the process improvement since an increment in the sulfate reduction at longer HRTs depends on the SRB activity and growth. In contrast, short HRTs could decrease the sulfide concentration as the sulfate-loading rate surpasses the sulfate reduction rate. Similar observations were encountered in a reactor operated at pH 4-6 with a pH auxostat in which the HRT solely depends on the sulfate conversion rate (Bijmans et al. 2008b; Bijmans et al. 2009).

7.5 Conclusions

• The statistical analysis carried out to the operational performance of the sulfate reducing IFB bioreactors revealed the clear roles played by the different process variables on the sulfate reduction process.

- The most significant variable affecting the process was the pH, whereas the other variables affected mainly the sulfide concentration response.

- The combined effect of the pH and COD concentration enhanced the sulfate reduction process, while the individual COD concentration effect only enhanced the COD removal efficiency, suggesting changes on the microbial pathway.

- The addition of metals was negative for the process, but the inhibition mechanism depended on the M/S^{2-} ratio.

References

APHA APHA (2005) Standard methods for examination of water and wastewater. 20 edn., Washington D.C.

Bijmans MFM, Buisman CJN, Meulepas RJW, Lens PNL (2011) 6.34 - Sulfate Reduction for Inorganic Waste and Process Water Treatment. In: Editor-in-Chief: Murray M-Y (ed) Comprehensive Biotechnology (Second Edition). Academic Press, Burlington, pp 435-446

Bijmans MFM, de Vries E, Yang C-H, N. Buisman CJ, Lens PNL, Dopson M (2010) Sulfate reduction at pH 4.0 for treatment of process and wastewaters. Biotechnology Progress 26 (4):1029-1037. doi:10.1002/btpr.400

Bijmans MFM, Dopson M, Ennin F, Lens PNL, Buisman CJN (2008a) Effect of sulfide removal on sulfate reduction at pH 5 in a hydrogen fed gas-lift bioreactor. Journal of microbiology and biotechnology 18 (11):1809-1818

Bijmans MFM, Peeters TWT, Lens PNL, Buisman CJN (2008b) High rate sulfate reduction at pH 6 in a pH-auxostat submerged membrane bioreactor fed with formate. Water Research 42 (10–11):2439-2448. doi:10.1016/j.watres.2008.01.025

Bijmans MFM, van Helvoort P-J, Buisman CJN, Lens PNL (2009) Effect of the sulfide concentration on zinc bio-precipitation in a single stage sulfidogenic bioreactor at pH 5.5. Separation and Purification Technology 69 (3):243-248

Canfield DE, Raiswell R (1999) The evolution of the sulfur cycle. American Journal of Science 299 (7-9):697-723. doi:10.2475/ajs.299.7-9.697

Cord-Ruwisch R (1985) A quick method for the determination of dissolved and precipitated sulfides in cultures of sulfate-reducing bacteria. Journal of Microbiological Methods 4 (1):33-36

Chen Y, Cheng JJ, Creamer KS (2008) Inhibition of anaerobic digestion process: A review. Bioresource Technology 99 (10):4044-4064. doi:10.1016/j.biortech.2007.01.057

Elferink S, Krooneman J, Gottschal JC, Spoelstra SF, Faber F, Driehuis F (2001) Anaerobic conversion of lactic acid to acetic acid and 1,2-propanediol by Lactobacillus buchneri. Applied and environmental microbiology 67, 125-132

Gallegos-Garcia M, Celis LB, Rangel-Méndez R, Razo-Flores E (2009) Precipitation and recovery of metal sulfides from metal containing acidic wastewater in a sulfidogenic down-flow fluidized bed reactor. Biotechnology and Bioengineering 102 (1):91-99

Gonzalez-Silva BM, Briones-Gallardo R, Razo-Flores E, Celis LB (2009) Inhibition of sulfate reduction by iron, cadmium and sulfide in granular sludge. Journal of Hazardous Materials 172 (1):400-407. doi:10.1016/j.jhazmat.2009.07.022

Huisman JL, Schouten G, Schultz C (2006) Biologically produced sulphide for purification of process streams, effluent treatment and recovery of metals in the metal and mining industry. Hydrometallurgy 83 (1-4):106-113

Johnson DB, Hallberg KB (2005) Biogeochemistry of the compost bioreactor components of a composite acid mine drainage passive remediation system. Science of The Total Environment 338 (1–2):81-93. doi:10.1016/j.scitotenv.2004.09.008

Kaksonen AH, Plumb JJ, Robertson WJ, Riekkola-Vanhanen M, Franzmann PD, Puhakka JA (2006) The performance, kinetics and microbiology of sulfidogenic fluidized-bed treatment of acidic metal- and sulfate-containing wastewater. Hydrometallurgy 83 (1-4):204-213

Kaksonen AH, Puhakka JA (2007) Sulfate Reduction Based Bioprocesses for the Treatment of Acid Mine Drainage and the Recovery of Metals. Engineering in Life Sciences 7 (6):541-564

Kimura S, Hallberg K, Johnson D (2006) Sulfidogenesis in Low pH (3.8–4.2) Media by a Mixed Population of Acidophilic Bacteria. Biodegradation 17 (2):57-65. doi:10.1007/s10532-005-3050-4

Labrenz M, Druschel GK, Thomsen-Ebert T, Gilbert B, Welch SA, Kemner KM, Logan GA, Summons RE, Stasio GD, Bond PL, Lai B, Kelly SD, Banfield JF (2000) Formation of Sphalerite (ZnS) Deposits in Natural Biofilms of Sulfate-Reducing Bacteria. Science 290 (5497):1744-1747. doi:10.1126/science.290.5497.1744

Lopes SIC, Sulistyawati I, Capela MI, Lens PNL (2007) Low pH (6, 5 and 4) sulfate reduction during the acidification of sucrose under thermophilic (55 °C) conditions. Process Biochemistry 42 (4):580-591. doi:10.1016/j.procbio.2006.11.004

Montgomery DC (2004) Design and Analysis of Experiments (7th Edition). John Wiley & Sons,

Moosa S, Harrison STL (2006) Product inhibition by sulphide species on biological sulphate reduction for the treatment of acid mine drainage. Hydrometallurgy 83 (1–4):214-222. doi:10.1016/j.hydromet.2006.03.026

Neculita C-M, Zagury GJ, Bussiere B (2007) Passive Treatment of Acid Mine Drainage in Bioreactors using Sulfate-Reducing Bacteria: Critical Review and Research Needs. J Environ Qual 36 (1):1-16. doi:10.2134/jeq2006.0066

Oyekola OO, van Hille RP, Harrison STL (2009) Study of anaerobic lactate metabolism under biosulphidogenic conditions. Water Research 43: 3345-3354

Papirio S, Villa-Gomez DK, Esposito G, Lens PNL, Pirozzi F (2012) Acid mine drainage treatment in fluidized-bed bioreactors by sulfate-reducing bacteria: a critical review. Critical reviews in environmental science and technology In Press

Reis MAM, Almeida JS, Lemos PC, Carrondo MJT (1992) Effect of hydrogen sulfide on growth of sulfate reducing bacteria. Biotechnology and Bioengineering 40 (5):593-600. doi:10.1002/bit.260400506

Tabak HH, Scharp R, Burckle J, Kawahara FK, Govind R (2003) Advances in biotreatment of acid mine drainage and biorecovery of metals: 1. Metal precipitation for recovery and recycle. Biodegradation 14 (6):423-436

Utgikar VP, Chaudhary N, Koeniger A, Tabak HH, Haines JR, Govind R (2004) Toxicity of metals and metal mixtures: analysis of concentration and time dependence for zinc and copper. Water Research 38 (17):3651-3658

Utgikar VP, Harmon SM, Chaudhary N, Tabak HH, Govind R, Haines JR (2002) Inhibition of sulfate-reducing bacteria by metal sulfide formation in bioremediation of acid mine drainage. Environmental Toxicology 17 (1):40-48

Villa-Gomez D, Ababneh H, Papirio S, Rousseau DPL, Lens PNL (2011) Effect of sulfide concentration on the location of the metal precipitates in inversed fluidized bed reactors. Journal of Hazardous Materials 192 (1):200-207. doi:10.1016/j.jhazmat.2011.05.002

Villa-Gomez D, Enright AM, E. L., A. B, Kramer H, P.N.L. L (2012a) Effect of hydraulic retention time on metal precipitation in sulfate reducing inverse fluidized bed reactors. Submited to Separation and Purification Technology

Villa-Gomez DK, Cassidy J, Keesman K, Sampaio R, P.N.L. L (2012b) Tuning strategies for sulfide control in sulfate reducing bioreactors using a pS electrode In preparation

White C, Gadd GM (1996) Mixed sulphate-reducing bacterial cultures for bioprecipitation of toxic metals: factorial and response-surface analysis of the effects of dilution rate, sulphate and substrate concentration. Microbiology 142 (8):2197-2205. doi:10.1099/13500872-142-8-2197

Zayed G, Winter J (2000) Inhibition of methane production from whey by heavy metals – protective effect of sulfide. Applied Microbiology and Biotechnology 53 (6):726-731. doi:10.1007/s002530000336

Zehnder AJB, Huser BA, Brock TD, Wuhrmann K (1980) Characterization of an acetate-decarboxylating, non-hydrogen-oxidizing methane bacterium. Archives of Microbiology 124 (1):1-11. doi:10.1007/bf00407022

8

GENERAL DISCUSSION

AND

RECOMMENDATIONS

8.1 Introduction

Metal-containing wastewaters represent an environmental and human health problem, but also a potential resource of valuable metals when their recovery is possible (Lens et al. 2002). These wastewaters typically contain high concentrations of dissolved metal-ions and sulfuric acid but are deficient in organic compounds (Johnson and Hallberg 2005; Papirio et al. 2012).

Several technologies have been applied to treat this type of wastewaters including chemical precipitation, ion exchange, adsorption, coagulation-flocculation and membrane filtration (Fu and Wang 2011). Not all these technologies are, however, suitable for metal recovery and may lead to high disposal expenses (Tabak et al. 2003; Esposito et al. 2006). Biological sulfate reduction is a process for the treatment of metal containing wastewaters enabling the recovery of metals as sulfidic precipitates (Bijmans et al. 2011). To date, metal recovery using this process has been demonstrated at large-scale when sulfate reduction and metal precipitation occur in separate stages (Huisman et al. 2006).

Simultaneous biogenic sulfide production and metal sulfide precipitation in a single unit simplifies the process design, reduces investment costs, and avoids transport of toxic sulfide to a precipitator reactor (Bijmans et al. 2009b). In this context, the inverse fluidized-bed reactor (IFB) is an attractive reactor configuration to allow recovery of metal sulfides at the bottom of the reactor, separated from the floating SRB biomass (Villa-Gomez et al. 2011; Gallegos-Garcia et al. 2009). Although metal removal has been achieved successfully in this reactor configuration (Chapter 2; Gallegos-Garcia et al., 2009), there are still some challenges to achieve metal recovery and practical implementation.

Metal recovery depends on the particle size of the metal sulfides, determined by the kinetics of nucleation, crystal growth and agglomeration (Al-Tarazi et al. 2004), as well as their purity and their dewatering characteristics. Several studies have correlated the size and morphology of the metal sulfide precipitates with the change in operational parameters such as pH, sulfide concentration (Sahinkaya et al. 2009; Sampaio et al. 2009; Veeken et al. 2003; Bijmans et al. 2009b; Al-Tarazi et al. 2005b) and reactor configuration (Al-Tarazi et al. 2005a). However, the bioreactor media contain biomass, organics and macronutrients supplied for bacterial growth that can also affect the metal precipitate characteristics for potential recovery (Esposito et al. 2006; Sampaio et al. 2009; Bijmans et al. 2009a). The study of the metal sulfide precipitate characteristics in bioreactors is not well documented, partly due to the lack of use of metal species-specific analytical methods for their identification (Neculita et al. 2007).

Practical implementation of metal sulfide precipitation and sulfate reduction in a single unit is also challenging because these metal containing waste streams are acid and contain metals that can affect the biological process performance. Several studies have reported the individual effect of different operational parameters such as the pH, HRT, electron donor concentration and metal to sulfide molar (M/S^{2-}) ratio in sulfate reducing bioreactors (Kaksonen and Puhakka 2007). However, the effect of key variables affecting the sulfate reduction process such as pH and metal addition connected to the reactor operational conditions (i.e. HRT, OLR) must be understood in a multi-variable context, and not only one variable-at-a-time effects.

Finally, these waste streams are deficient in organic compounds as electron donor for sulfate reduction. Thus, for practical implementation, steering the sulfide production towards its required stoichiometric amount in bioreactors is highly relevant to avoid unnecessary electron donor addition and over production of sulfide. Therefore, process control in sulfate reducing bioreactors is essential for the control of the sulfide concentration depending on the metal amount desired to precipitate. Mathematical models have been developed to support the design of a control strategy for sulfate reduction in bioreactors (Kalyuzhnyi and Fedorovich 1998; Oyekola et al. 2012; Gupta et al. 1994). However, in these studies the objective was to outcompete or favor microbial trophic groups other than SRB, while accounting for the control of the sulfide production.

8.2 Objective

The main objective of this thesis was to elucidate the factors affecting simultaneous sulfate reduction and precipitation of heavy metals in an inverse fluidized bed reactor (IFB) reactor in order to optimize the metal recovery from wastewaters such as acid mine drainage (AMD). Therefore, this thesis focused on varying different operational conditions to study their effect on the solid-liquid separation and purity of the metal sulfide precipitates as well as on their effect on the sulfate reducing process for process control (Figures 8.1, 8.2 and 8.3). In addition, recommendations for further research to improve the recovery of the metal sulfides in bioreactors are given.

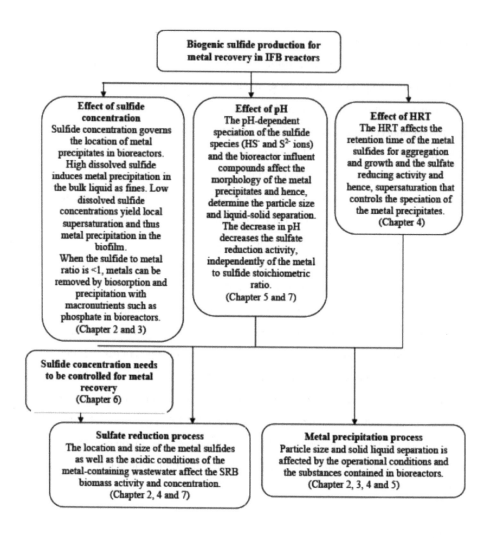

Figure 8.1. Overview of the structure of this thesis on the aspects of biogenic sulfide production for metal recovery in inverse fluidized bed reactors. HRT: hydraulic retention time; SRB: Sulfate reducing bacteria.

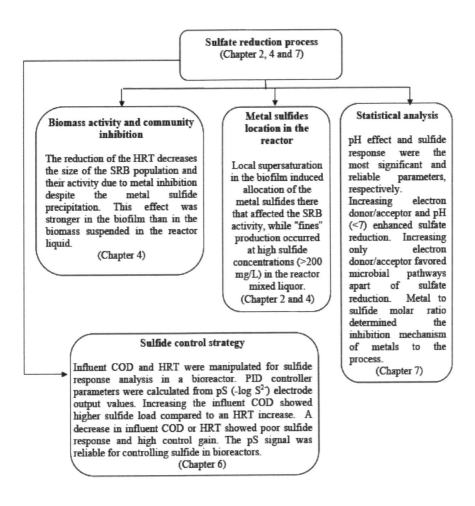

Figure 8.2. Overview of the results of this thesis concerning biomass activity and SRB community dynamics, metal sulfide allocation in the IFB reactor and metal removal mechanisms in the sulfate reducing process. HRT: hydraulic retention time; COD: chemical oxygen demand; SRB: sulfate reducing bacteria and PID: proportional integral derivative.

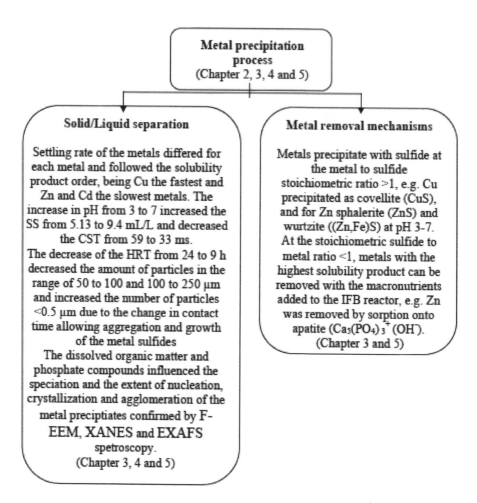

Figure 8.3. Overview of the results of this thesis concerning solid/liquid separation and metal removal mechanisms in the metal precipitation process. F-EEM: fluorescence excitation emission matrix; XANES: X-ray absorption near edge spectroscopy; EXAFS: extended X-ray absorption fine structure; SS: settleable solids and CST: capillary suction time.

8.3 Effect of sulfide, pH and HRT on the metal sulfide precipitation characteristics and sulfate reducing activity in the IFB reactor

This thesis shows that the characteristics of the metal sulfides formed in the IFB reactor depend on the operational conditions, such as the sulfide concentration (Chapter 1 and 2), the hydraulic retention time (HRT; Chapter 4) and the pH (Chapter 5). These parameters control the extent of agglomeration for particle growth and settling as well as the supersaturation for crystallization of the metal sulfides (Figure 3, Chapter 5).

8.3.1 Effect of sulfide concentration

8.3.1.1 Sulfide concentration affecting the location of metal sulfide precipitation in the IFB reactor

The sulfide concentration controlled the location of the metal sulfides in the IFB reactor. At high sulfide concentrations (>200 mg/L), high levels of supersaturation were present in the bulk liquid allowing the formation of "fines", while local supersaturation in the biofilm occurred at low sulfide concentrations (<50 mg/L) (Chapter 2), because biofilms function as nucleation seeds, enhancing the formation and growth of the metal sulfides crystals (Bijmans et al. 2009a).

Although the sulfide concentration ensures prior metal sulfide precipitation that prevented free metal toxicity (Kaksonen and Puhakka 2007), and that biofilms are reported to be more resistant towards high metal concentrations in comparison with bacterial cultures growing in suspension (van Hullebusch et al. 2003), in chapter 4 it was shown that the tendency of the metal sulfides to precipitate within/near the biofilm (Chapter 2) causes inhibition of the SRB activity and decreases their population size (Figure 6, Chapter 4) as previously reported (Utgikar et al. 2002).

8.3.1.2 Alternative metal removal mechanisms at low sulfide concentrations

The metals with the highest solubility product are not only removed by metal sulfide precipitation, but through a variety of removal mechanisms, particularly when the sulfide concentration is below the sulfide to metal stoichiometry (Figure 1, Chapter 3). These include biosorption, precipitation with macronutrients as phosphate and sorption onto their precipitates, e.g. apatite. Previous studies confirm the presence of alternative precipitates at low sulfide concentrations (Mokone et al. 2010; Neculita et al. 2007). However, these studies could not identify clearly the metal removal mechanisms because of the poor crystallinity of the metal precipitates. The use of species-specific analytical methods such as EXAFS and XANES, allowed to identify that Zn is also sorbed onto apatite $(Ca_5(PO_4)_3{}^+(OH^-)$ or hydroxyapatite (depending on the pH) and that these sorption mechanisms contribute to the Zn removal in bioreactors (Figure 4 and 5, Chapter 3).

8.3.2 Effect of HRT

The change in HRT affected the retention time of the metal sulfides for particle growth as well as the sulfate reducing activity (sulfide production) for metal sulfide precipitation (Chapter 4). The change in HRT from 24 to 9 h affected the size of the metal sulfide precipitates due to the change in contact time allowing aggregation and growth of the metal sulfides. The further HRT decrease to 4.5 h affected the sulfate reducing activity for sulfide production and hence, the supersaturation level. In addition, free metal toxicity caused process failure. Although the decrease in supersaturation is favorable for crystal growth of metal sulfides (Mersmann 1999), the formation of alternative precipitates such as brochantite $(Cu_4(OH)_6SO_4)$ (Mokone et al. 2010) and Zn-phosphate (Chapter 3) also occurred (Table 2, Chapter 4).

8.3.3 Effect of pH

The pH varies the influence of the influent macronutrients and the dissolved organic matter (DOM), present in the bioreactor, on the characteristics of the metal sulfide precipitates. At pH 5, the DOM from the bioreactor liquor (Figure 6, Chapter 5), induces crystallization and hampers agglomeration of the metal sulfides. This is explained as a passive binding between the metals and the organic matter that decreases the supersaturation of the metal sulfide precipitation and hence, allows crystallization over nucleation (Gadd et al., 2009; Johnson et al., 2007). In contrast, the metal sulfides formed at pH 7 showed aggregation of particles impeding the visualization of crystals via SEM (Figure 3, Chapter 5). In addition, the metal precipitates formed at pH 7 displayed faster settling rates and capillary suction times as well as a higher concentration of settleable solids. The reason could be that at this pH, the precipitates are also composed of phosphate (hydroxyapatite) (Figure 1b, Chapter 5) that are known to exhibit better liquid-solid separation as compared with the metal sulfides (Chapter 5, Djedidi et al., (2009).

The X-ray absorption fine structure spectroscopy analyses confirmed the precipitation of Zn-S as sphalerite and Cu-S as covellite in all the samples, but also revealed the presence of minor amounts of Zn-sorbed on hydroxyapatite. The analyses further showed a more organized ZnS structure at pH 5, in agreement with the scanning electron microscopy images of cubic sphalerite type structures. In contrast, amorphous CuS structures were observed regardless the pH.

8.4 Multivariable analysis of the effect of the operational conditions on the sulfate reducing process

A statistical analysis yielded valuable information on the synergistic and antagonistic effects of the influent pH, chemical oxygen demand concentration (COD_{in}), metal to sulfide (M/S^{2-}) ratio and HRT on the sulfate reduction process (Chapter 7). The three variables evaluated (COD, sulfate removal efficiency, and sulfide concentration) evidenced the strong effect of the pH over the other operational parameters (Figure 1 and 3, Chapter 7). The pH as well as the COD_{in} can lead to a selection of microbial species that was reflected in the differences between COD removal and sulfate reduction effiency (Figure 4). This is because an excess of COD_{in} enhances the biomass growth rate and diversity of microbial species (Oyekola et al., 2009) that can help to maintain sulfate reduction at adverse conditions such as low pH (Elferink et al. 2001). The RSA showed that the M/S^{2-} ratio determined whether the inhibition mechanism to the SR process was due to the presence of free metals (Figure 5 right, Chapter 7) or precipitated metal sulfides (Utgikar et al. 2002; Utgikar et al. 2004).

8.5 Control and monitoring of the sulfide concentration in the IFB reactor

This Ph.D study shows that the control of the sulfide concentration in the bioreactor liquid for metal sulfide precipitation can be applied to avoid electron donor losses, to obtain pure metal sulfides (Chapter 2, 4, and 5) and to avoid biomass toxicity (Chapter 4). Therefore, tuning strategies to control the sulfide concentration in sulfate reducing bioreactors were evaluated using a direct measurement of the sulfide species (pS) for the acquisition of proportional-integral-derivative (PID) controller parameters (Chapter 6). The PID parameters from the different step response experiments with both tuning strategies (Table 2) provided a baseline to evaluate the potential of the PID controller for its application in SRB bioreactors and to foresee the parameters that need to be adapted or included (Heinzle et al. 1993).

The changes in the OLR during the IFB reactor operation show rapid responses in the sulfide concentration and a higher sulfide increment when a change of COD_{in}

177

concentration was applied compared to the change in the OLR via the manipulation of the HRT. The decrease of the sulfide concentration was more difficult to achieve as it displayed a longer response time irrespective of the manipulated variable (COD_{in} and HRT) (Figure 5, Chapter 6). Therefore, another strategy to decrease the sulfide concentration in the IFB bioreactor should be considered such as the dilution of the effluent, as used in the control of wastewater treatment plants (Metcalf and Eddy 2002). Delays in the response time and a high control gain were the most critical factors affecting the application of a sulfide control strategy in bioreactors. These were caused by the induction of different metabolic pathways in the anaerobic sludge. These included substrate utilization by microbial groups other than SRB such as lactate fermentation to acetate (Oyekola et al. 2012) when the HRT was manipulated (Table 1, Chapter 6) and the accumulation of storage products under feast-famine conditions (Hai et al. 2004) promoted with the increase and decrease of the OLR via the manipulation of the COD_{in}.

In addition, chapter 6 showed that the online measurement of the sulfide concentration via the pS (Figure 2 and 3, Chapter 6) can be used for the control of the sulfide concentration in SRB bioreactors. However, pH variations and high sulfide concentrations should be carefully observed for correction of the recorded pS values.

8.6 Implication for biotechnological applications

The small particle size of the metal sulfides (Chapter 2, 3, 4 and 5) and the adaptation of the controller parameters for sulfide control (Chapter 5) are an important bottleneck towards the biotechnological application of the IFB sulfate reducing reactor for metal recovery in AMD waste streams.

At sulfide concentrations above the sulfide to metal stoichiometric ratio, high supersaturation conditions are difficult to be avoided (Chapter 1 and 2) and thus, crystal growth is less relevant for particle growth as compared to the agglomeration of the metal sulfide precipitates (Figure 6, Chapter 2). As agglomerates are, however, relatively fragile, the precipitates displayed a wide range in size distribution (0.48-250 μm; Chapter 4) that varied with the retention time of the crystals in the IFB reactor (Chapter 4). Thus, promoting agglomeration for particle growth might be a difficult option to control the particle size compared with the manipulation of other parameters that favour the formation of more compact crystals.

Crystal growth can be favoured at low sulfide concentrations and hence, the decrease of the supersaturation in the bioreactor (Mersmann 1999). However, up to now, the maximum crystal size for several metal sulfides is reported to be below 10 μm (Lewis 2010). Such particle sizes are still difficult to settle using standard clarifiers for solid state separation (WEF, 2005). Besides, low sulfide concentrations bring a tradeoff between crystallization and metal sulfide purity (Chapter 3 and 4).

The evaluation of tuning strategies to control the sulfide concentration was the first approach to obtain the PID parameters towards an automated metal precipitation/recovery in the IFB reactor (Chapter 6). However, several issues were encountered related to variable process dynamics that could affect the stability of the process control such as delays in response time and a high control gain (Table 2, Chapter 6), which may cause an excessive control action and instabilities in the reactor (Stephanopoulos 1984). These issues have been observed when PID controllers are used in biological systems, since time delays cannot be incorporated in the system model (Mailleret et al. 2004; Alvarez-Ramirez et al. 2002). In addition, the accumulation of storage products allows the SRB to continue producing sulfide even when the organic loading rate is decreased, thus causing an unexpected sulfide response that cannot be considered in the conventional PID controller. Therefore, the adaptation and introduction of additional parameters to the controller should be explored.

8.7 Recommendations for future research

8.7.1 Seeding material to promote crystallization

In chemical precipitators, "seeding material" are used to promote crystal growth and thus to increase the size of the precipitates (Mersmann 1999). Several authors have induced the precipitation of heavy metals on the sand surface in fluidized bed reactors using sulfide or carbonate (Zhou et al., 1991; Guillard et al 2001; van Hille et al., 2005). Zhou et al. (1991) observed that when the ratio of carbonate to metals (Cu, Ni and Zn) was low, metal precipitates were coated on the sand surface, while at high ratios (6:1 and 3:1) the precipitation occurred in the bulk solution. Similarly, this study shows that the metal sulfides accumulated in the polyethylene beads coated with SRB biofilm till the point where these beads became so heavy that they lost their floating characteristic (Chapter 2). Even though, the crystallization characteristics of the metal sulfides precipitated in the biofilm was not determined in this work, such biomass could be an alternative not only to enhance crystal growth but also to separate the metals from the liquid phase. However, the metal precipitates are then impure and biomass will be lost from the bioreactor upon metal harvesting, which is undesirable as the growth rate of SRB is rather slow (Widdel, 1988). To overcome this, a strategy could be explored to separate and re-circulate the biomass back to the bioreactor after the metal sulfides have been recovered.

8.7.2 Alternative substrates promoting EPS that can decrease supersaturation

In chapter 5, it was observed that the DOM produced under the IFB reactor operational conditions helped to mediate the crystallization process of metal sulfides probably by reducing the supersaturation at pH 5. Several studies have examined the influence of the operational conditions and type of substrate on the characteristics of the extracellular polymeric substances (EPS) produced to understand biofouling in membrane bioreactors (Wu et al. 2012). Wu et al. (2012) found that a fulvic acid-like component was the main responsible for Cd(II) binding while Cu(II) inclined to complex with humic-like components rather than protein-like ones. Additionally, the molecular weight of the substrate analyzed was found to exert less influence on metal binding than that of specific metals. In this Ph.D study, the changes in fluoresecence intensity of the DOM are related to the use of lactate as substrate at one single concentration (Table 2, Chapter 5). Other types or different concentrations of substrate could lead to different characteristics of the DOM affecting the metal sulfide precipitate characteristics. Therefore, a deeper study on the characteristics of the DOM produced in sulfate reducing reactors for the enhancement of crystallization should be done.

8.7.3 Additional settlers for sequential precipitation

The degree of agglomeration for particle growth and settling on the metal sulfides was shown to be metal specific and linked to their solubility product (Chapters 3, 5 and 5). During the reactor operation (Chapter 4) and the metal depletion kinetics in the batch experiments (Chapter 2 and 3), Cd and Zn had lower settling rates compared to Cu and Pb. This study (Chapter 3), in agreement with a previous study (Al-Tarazi et al. 2005b), has shown that the agglomeration of precipitates is correlated with the solubility product (Blais et al. 2008). This is because nucleation enhances agglomeration, and the nucleation rate is dependent on the solubility product (Al-Tarazi et al. 2005b). This characteristic can be used for the selective recovery of the metal with the lowest solubility product from a multi-metal wastewater in the IFB reactor itself acting as the first settler, whereas the remaining metal sulfides with higher solubility product can be recovered in additional settlers placed in sequence with the IFB reactor.

8.7.4 Adaptation of the PID controller for the control of the sulfide concentration

Delays in response time and a high control gain (Table 2, Chapter 6) were found the most critical factors affecting the application of a control strategy in bioreactors. Therefore, the PID controller should be modified to reduce these factors. One alternative can be to decrease the timing of the output (Steyer et al. 2000), in this case, the pS readings. However, this option might cause information losses of significant changes affecting the bioreactor, and consequently, operation failures that could be prevented with the pS readings information.

The use of an adaptive PID controller with gain scheduling or a model-based adaptive controller (Haugen, 2004) could be a better option to overcome delays in response time and to diminish a high control gain both caused by variations in the SRB activity. An approach for the design of linear feedback controllers for anaerobic digestion systems was presented by Alvarez-Ramirez et al. (2002) using a PI controller with a novel control gain configuration to regulate the effluent COD concentration. They proposed a cascade control structure to overcome time delays by measuring VFA concentrations instead of the total COD concentration. In the case of the use of a model-based adaptive controller, the parameters are estimated continuously based on the changes in the microbial activity and population (Rodrigo et al., 1999). This information can be obtained by the dynamics of the bioprocess such as reaction pathways and mass balances (Steyer et al. 2000).

References

Al-Tarazi M, Heesink ABM, Azzam MOJ, Yahya SA, Versteeg GF (2004) Crystallization kinetics of ZnS precipitation; an experimental study using the mixed-suspension-mixed-product-removal (MSMPR) method. Crystal Research and Technology 39 (8):675-685. doi:10.1002/crat.200310238

Al-Tarazi M, Heesink ABM, Versteeg GF (2005a) Effects of reactor type and mass transfer on the morphology of CuS and ZnS crystals. Crystal Research and Technology 40 (8):735-740

Al-Tarazi M, Heesink ABM, Versteeg GF, Azzam MOJ, Azzam K (2005b) Precipitation of CuS and ZnS in a bubble column reactor. AIChE Journal 51 (1):235-246

Alvarez-Ramirez J, Meraz M, Monroy O, Velasco A (2002) Feedback control design for an anaerobic digestion process. Journal of Chemical Technology & Biotechnology 77 (6):725-734. doi:10.1002/jctb.609

Bijmans MFM, Buisman CJN, Meulepas RJW, Lens PNL (2011) 6.34 - Sulfate Reduction for Inorganic Waste and Process Water Treatment. In: Editor-in-Chief: Murray M-Y (ed) Comprehensive Biotechnology (Second Edition). Academic Press, Burlington, pp 435-446

Bijmans MFM, van Helvoort P-J, Buisman CJN, Lens PNL (2009a) Effect of the sulfide concentration on zinc bio-precipitation in a single stage sulfidogenic bioreactor at pH 5.5. Separation and Purification Technology 69 (3):243-248

Bijmans MFM, van Helvoort P-J, Dar SA, Dopson M, Lens PNL, Buisman CJN (2009b) Selective recovery of nickel over iron from a nickel-iron solution using microbial sulfate reduction in a gas-lift bioreactor. Water Research 43 (3):853-861

Blais J, Djedidi Z, Cheikh R, Tyagi R, Mercier G (2008) Metals Precipitation from Effluents: Review. Practice Periodical of Hazardous, Toxic, and Radioactive Waste Management 12 (3):135-149. doi:doi:10.1061/(ASCE)1090-025X(2008)12:3(135)

Djedidi Z, Khaled JB, Cheikh RB, Blais J-F, Mercier G, Tyagi RD (2009) Comparative study of dewatering characteristics of metal precipitates generated during treatment of monometallic solutions. Hydrometallurgy 95 (1-2):61-69. doi:10.1016/j.hydromet.2008.04.014

Elferink S, Krooneman J, Gottschal JC, Spoelstra SF, Faber F, Driehuis F (2001) Anaerobic conversion of lactic acid to acetic acid and 1,2-propanediol by Lactobacillus buchneri. Applied and environmental microbiology 67, 125-132

Esposito G, Veeken A, Weijma J, Lens PNL (2006) Use of biogenic sulfide for ZnS precipitation. Separation and Purification Technology 51 (1):31-39

Fu F, Wang Q (2011) Removal of heavy metal ions from wastewaters: A review. Journal of Environmental Management 92 (3):407-418. doi:10.1016/j.jenvman.2010.11.011

Gallegos-Garcia M, Celis LB, Rangel-Méndez R, Razo-Flores E (2009) Precipitation and recovery of metal sulfides from metal containing acidic wastewater in a sulfidogenic down-flow fluidized bed reactor. Biotechnology and Bioengineering 102 (1):91-99

Gupta A, Flora JRV, Sayles GD, Suidan MT (1994) Methanogenesis and sulfate reduction in chemostats—II. Model development and verification. Water Research 28 (4):795-803. doi:10.1016/0043-1354(94)90086-8

Hai T, Lange D, Rabus R, Steinbüchel A (2004) Polyhydroxyalkanoate (PHA) accumulation in sulfate-reducing bacteria and identification of a class III PHA synthase (PhaEC) in Desulfococcus multivorans. Appl Environ Microbiol 70 (8):4440-4448

Heinzle E, Dunn I, Ryhiner G (1993) Modeling and control for anaerobic wastewater treatment

Bioprocess Design and Control. In, vol 48. Advances in Biochemical Engineering/Biotechnology. Springer Berlin / Heidelberg, pp 79-114. doi:10.1007/BFb0007197

Huisman JL, Schouten G, Schultz C (2006) Biologically produced sulphide for purification of process streams, effluent treatment and recovery of metals in the metal and mining industry. Hydrometallurgy 83 (1-4):106-113

Johnson DB, Hallberg KB (2005) Biogeochemistry of the compost bioreactor components of a composite acid mine drainage passive remediation system. Science of The Total Environment 338 (1–2):81-93. doi:10.1016/j.scitotenv.2004.09.008

Kaksonen AH, Puhakka JA (2007) Sulfate Reduction Based Bioprocesses for the Treatment of Acid Mine Drainage and the Recovery of Metals. Engineering in Life Sciences 7 (6):541-564

Kalyuzhnyi SV, Fedorovich VV (1998) Mathematical modelling of competition between sulphate reduction and methanogenesis in anaerobic reactors. Bioresource Technology 65 (3):227-242. doi:10.1016/s0960-8524(98)00019-4

Lens PNL, Hulshoff Pol L, Wilderer P (2002) Water Recycling and Resource Recovery in Industry: Analysis, Technologies and Implementation. Integrated environmental Technology Series. IWA Publishing, Cornwall

Lewis AE (2010) Review of metal sulphide precipitation. Hydrometallurgy 104 (2):222-234

Mailleret L, Bernard O, Steyer JP (2004) Nonlinear adaptive control for bioreactors with unknown kinetics. Automatica 40 (8):1379-1385. doi:10.1016/j.automatica.2004.01.030

Mersmann A (1999) Crystallization and precipitation. Chemical Engineering and Processing 38 (4-6):345-353

Metcalf, Eddy (2002) Wastewater Enginering: Treatmt & Reuse. McGraw-Hill Education (India) Pvt Limited,

Mokone TP, van Hille RP, Lewis AE (2010) Effect of solution chemistry on particle characteristics during metal sulfide precipitation. Journal of Colloid and Interface Science 351 (1):10-18. doi:10.1016/j.jcis.2010.06.027

Neculita C-M, Zagury GJ, Bussiere B (2007) Passive Treatment of Acid Mine Drainage in Bioreactors using Sulfate-Reducing Bacteria: Critical Review and Research Needs. J Environ Qual 36 (1):1-16. doi:10.2134/jeq2006.0066

Oyekola OO, Harrison STL, van Hille RP (2012) Effect of culture conditions on the competitive interaction between lactate oxidizers and fermenters in a biological sulfate reduction system. Bioresource Technology 104 (0):616-621. doi:10.1016/j.biortech.2011.11.052

Papirio S, Villa-Gomez DK, Esposito G, Lens PNL, Pirozzi F (2012) Acid mine drainage treatment in fluidized-bed bioreactors by sulfate-reducing bacteria: a critical review. Critical reviews in environmental science and technology In Press

Sahinkaya E, Gungor M, Bayrakdar A, Yucesoy Z, Uyanik S (2009) Separate recovery of copper and zinc from acid mine drainage using biogenic sulfide. Journal of Hazardous Materials 171 (1-3):901-906

Sampaio RMM, Timmers RA, Xu Y, Keesman KJ, Lens PNL (2009) Selective precipitation of Cu from Zn in a pS controlled continuously stirred tank reactor. Journal of Hazardous Materials 165 (1-3):256-265

Stephanopoulos G (1984) Chemical process control. Prentice-Hall international series in the physical and chemical engineering sciences Prentice-Hall

Steyer JP, Bernet N, Lens PNL, Moletta R (2000) Anaerobic treatment of sulfate rich wastewaters : process modeling and control. In: In: Environmental Technologies to Treat Sulfur Pollution / P.N.L. Lens and L.W. Hulshoff Pol. - London : IWA Publishing, 2000. - ISBN 1900222094. pp 207-235

Tabak HH, Scharp R, Burckle J, Kawahara FK, Govind R (2003) Advances in biotreatment of acid mine drainage and biorecovery of metals: 1. Metal precipitation for recovery and recycle. Biodegradation 14 (6):423-436

Utgikar VP, Chaudhary N, Koeniger A, Tabak HH, Haines JR, Govind R (2004) Toxicity of metals and metal mixtures: analysis of concentration and time dependence for zinc and copper. Water Research 38 (17):3651-3658

Utgikar VP, Harmon SM, Chaudhary N, Tabak HH, Govind R, Haines JR (2002) Inhibition of sulfate-reducing bacteria by metal sulfide formation in bioremediation of acid mine drainage. Environmental Toxicology 17 (1):40-48

van Hullebusch E, Zandvoort M, Lens P (2003) Metal immobilisation by biofilms: Mechanisms and analytical tools. Reviews in Environmental Science and Biotechnology 2 (1):9-33

Veeken AHM, de Vries S, van der Mark A, Rulkens WH (2003) Selective Precipitation of Heavy Metals as Controlled by a Sulfide-Selective Electrode. Separation Science and Technology 38 (1):1 - 19

Villa-Gomez D, Ababneh H, Papirio S, Rousseau DPL, Lens PNL (2011) Effect of sulfide concentration on the location of the metal precipitates in inversed fluidized bed reactors. Journal of Hazardous Materials 192 (1):200-207. doi:10.1016/j.jhazmat.2011.05.002

Wu J, Zhang H, Shao L-M, He P-J (2012) Fluorescent characteristics and metal binding properties of individual molecular weight fractions in municipal solid waste leachate. Environmental Pollution 162 (0):63-71. doi:10.1016/j.envpol.2011.10.017

Summary

This study shows that the contribution of nucleation, crystal growth and agglomeration on the particle size and purity of the metal sulfides in the inverse fluidized bed (IFB) reactor varies with the sulfide concentration, the residence time of the precipitates and the pH.

The sulfide concentration was found to control the location of the metal sulfides in the IFB reactor as well as their purity, but not the particle size. At high sulfide concentrations (>200 mg/L), high levels of supersaturation were in the bulk liquid allowing the formation of "fines", while local supersaturation in the biofilm occurred at low sulfide concentrations (<50 mg/L). The allocation of the metal sulfides in the biofilm affects its SRB activity, while this was not observed in the SRB suspended in the bioreactor liquid. Metal removal mechanisms other than metal sulfide precipitation were favored when the sulfide concentration was below the sulfide to metal stroichiometric ratio. These included biosorption, precipitation with macronutrients as phosphate and sorption onto their precipitates, e.g. apatite.

The change in HRT affected the retention time of the metal sulfides for particle growth and the sulfate reducing activity for sulfide production. As a consequence, variations in the size of the metal sulfide precipitates were observed. The decrease of the HRT also affected the SRB activity and quantity that was partly attributed to the variations of the size of the metal sulfide precipitates but also to the insufficient sulfide concentration that allowed free metal toxicity.

The pH was found to determine the influence the dissolved organic matter (DOM) and the compounds present in the biogenic sulfide such as phosphate, on the characteristics of the metal sulfide precipitates. At pH 5, the DOM present in the bioreactor liquor, induced crystallization and hampered agglomeration of the metal sulfides due to a passive binding mechanism between the metals and the -COOH and –OH groups that decreases the supersaturation. In contrast, the metal sulfides formed at pH 7 showed aggregation of particles, faster settling rates, more settleable solids and better dewaterability. This as suspected to be due to the presence of phosphate precipitates as $Zn_3(PO_4)_2*4H_2O$ and hydroxyapatite that are known to exhibit better liquid-solid separation as compared with metal sulfides.

A statistical analysis was made with the results of the IFB reactor runs operating at different influent pH, chemical oxygen demand concentration (COD_{in}), metal to sulfide (M/S^{2-}) ratio and HRT to gain knowledge on their individual and combined effects of these parameters on the sulfate reduction process. Based on the results of analysis of variance (ANOVA) and response surface analysis, the combined effect of pH (up to 7) and COD_{in} (up to 3 gCOD/L) was positive for sulfate reduction, whereas the individual effect of COD_{in} was not significant at all. The effect of metal addition on the process performance depended on the M/S^{2-} ratio. The COD removal and sulfate reducing process were adversely affected with the addition of metals (Zn, Pb, Cu, Cd) in the influent, however, the effect differed and highly depended on the M/S^{2-} ratio as it determines insoluble metals sulfide inhibition and free metal toxicity.

As a result of the general conclusion about the need to control the sulfide concentration for process improvement in the previous chapters, this study evaluated two tuning strategies to manipulate the OLR via changing the COD_{in} or the HRT for controlling the sulfide concentration in sulfate reducing bioreactors for acquisition of the proportional-integral-derivative (PID) controller parameters. The changes in the OLR during the IFB reactor operation showed rapid responses in the sulfide concentration when the change was applied to the COD_{in} and a higher sulfide increment compared to the change in the OLR via the manipulation of the HRT. The step change applied by the manipulation of either COD_{in} or HRT for sulfide decrease displayed a small change and a longer response time. Changes in microbial activity affected the sulfide concentration set-point after the changes in the OLR. Thus, the information obtained by the dynamics of the bioprocess or an adaptive PID controller that includes the modeling of the system should be used for constant updating of the PID parameters.

Samenvatting

Deze studie toont aan dat in IFB (inverse fluidized bed) reactoren de sulfideconcentratie, de verblijftijd van de sulfideneerslagen en de pH bijdroegen aan het ontstaan en de groei van metaalsulfide kristallen en aan de zuiverheid ervan. De sulfideconcentratie bepaalde the plaats van de metaalsulfiden in de IFB reactor: hoge sulfideconcentraties (>200 mg/L) leidden tot superverzadiging in de hele reactor en de vorming van kleinere deeltjes, terwijl lokale superverzadiging in de biofilm optrad bij lage sulfideconcentraties (<50 mg/L). De activiteit van de sulfaatreducerende bacteriën in de biofilm werd beïnvloed door de plaats van de metaalsulfiden in de biofilm; uiteraard gold dit niet voor de activiteit van de bacteriën gesuspendeerd in bioreactor. Als de sulfideconcentratie onder de stoichiometrische verhouding sulfide-metaal kwam, kregen andere metaalverwijderende mechanismen dan het neerslaan van metaalsulfide de overhand, zoals biosorptie, neerslag met macronutriënten (fosfaat) en adsorptie aan apatiet.

Verandering van de HRT (hydraulische verblijftijd) beïnvloedde niet alleen de verblijftijd van de metaaldeeltjes en dus hun aangroei, maar ook de sulfaatreducerende activiteit en dus de sulfideproductie. Dientengevolge was er variatie in de grootte van de metaalsulfide deeltjes. De verlaging van de SRB activiteit bij verlaagde HRT kon deels worden toegeschreven aan de veranderde deeltjesgrootte van de metaalsulfiden, maar ook aan verhoogde toxiciteit van vrije metalen, die niet konden neerslaan bij gebrek aan voldoende sulfide.

Bij pH 5 stimuleerde het aanwezige DOM (opgeloste organische stof) de kristallisatie, maar verhinderde de samenvoeging van afzonderlijke metaalsulfidedeeltjes. De metaalsulfides gevormd bij pH 7 daarentegen aggregeerden wel en bezonken sneller. Ook hun ontwaterbaarheid was groter door de aanwezige fosfaatneerslagen, zoals $Zn_3(PO_4)_2.4H_2O$ en hydroxyapatiet.

Met behulp van een statistische analyse (ANOVA) zijn zowel de afzonderlijke als de gecombineerde effecten van pH, CZV_{in}, metaal/sulfide ratio en HRT op de sulfaatreductie in kaart gebracht. De combinatie van verhoging van pH (tot 7) en CZV (tot 3g/L) stimuleerde de sulfaatreductie, terwijl alleen CZV verhoging geen effect had. Het effect van metaaltoevoeging op het proces hing vooral af van de verhouding M/S^{2-}. De CZV verwijdering en de sulfaatreductie activiteit waren omgekeerd evenredig met de toevoeging van metalen (Zn, Pb, Cu, Cd), hoewel het effect verschilde en vooral afhankelijk leek te zijn

187

van de M/S^{2-} verhouding, omdat dit de remming door vrije, niet neergeslagen metalen beïnvloedt.

De discussie in de eerste hoofdstukken over de noodzaak om de sulfideconcentratie te kunnen reguleren ter verbetering van het proces, leidde tot twee "tuning strategies" om de organische belasting te kunnen manipuleren: hetzij via de CZV$_{in}$, hetzij via de HRT zou de sulfideconcentratie in sulfaatreducerende reactoren gecontroleerd kunnen worden. Dit alles om de juiste parameters te kunnen krijgen voor de PID (proportional-integral-derivative) controller.

Als de organische belasting in de IFB reactoren werd aangepast door middel van de CZV$_{in}$, dan wisselde de sulfideconcentratie sneller en met grotere sprongen dan wanneer die aanpassing kwam door verandering van de HRT. Stapsgewijze sulfide verlaging door het manipuleren van hetzij de CZV$_{in}$, hetzij de HRT, leidde tot kleinere veranderingen en een langere responstijd.

Door veranderingen in de organische belasting en de daarmee gepaard gaande veranderingen in microbiologische activiteit veranderde het sulfideconcentratie set-point. De PID parameters moeten dan ook continu worden aangepast, hetzij op grond van resultaten verkregen uit het proces, hetzij door middel van een adaptieve, modelgestuurde PID controller.

Acknowledgments

After all the effort, it is a pleasured to thank all the people who intervened during my PhD formation and my life in general during these four years and even before that. To do a PhD in the Netherlands was an experience that marked my life both personally and professionally. So I'll start by thanking destiny or God for guiding my life to end up studying in such a beautiful country. Secondly, I thank the National Council for Science and Technology (CONACYT, Mexico), which financed my research at UNESCO-IHE. I know that the opportunities in my country are getting more difficult so I thank them for believe in me.

A special mention goes to my supervisor Piet Lens, thank you for all the knowledge that you helped me to acquire. With all the discussions, papers, presentations, deadlines etc., you helped me to be open minded, keen observed, to look at situations from many angles, not being frustrated in finding one or several plausible solutions regardless of the time involved, and to use failure to improve future approaches to problem solving.

I would also thank Eric van Hullebusch for all the discussions and hard work at the European Syncrotron Radiation Facility, I appreciate your support on the metal precipitate characteristics analysis.

Thanks to Francois Farges for the nice discussions and for showing me those nice things about Mexico that I did not know. Thanks to Herman Kramer for the useful discussions on crystallization.

I thank also Henk Luberding, who despite of not being part of my committee he always helped me with questions about any topic and encouraged me to be positive during difficulties on my work, in particular towards the end of my PhD period.

A warm thanks to everybody at the Pollution Prevention and Control group as well as the Environmental Engineering and Water Technology department, also to all the technicians in the lab for their help and support.

To all the students, MSc students, PhD students, postdocs, staff members and all the people that contribute to create a great and multicultural atmosphere at UNESCO-IHE.

Many thanks to all the students that I mentored and helped me with my PhD research: Hisham Ababneh, Stephano Papirio, Rafaella Maestro, Eky Lystia, Samson Mushi, Joana Cassidy, Audrey Buttice and Kristen. Not only you helped me with my PhD work but also to acquire social (multicultural) and teaching skills that I am sure going to use in the future.

I want to thank Anne Marie Enright for the microbial population analysis from chapter 4.

I special thank to my Msc supervisor Elias Razo for putting the seed on me about doing a PhD in The Netherlands, thanks for believe in me.

During my first PhD year I had the pleasure to have as colleagues to Roel Meulepas and Jan Bartacek, thank you for all you support in the beginning. Roel thanks for all the help and for showing me your beautiful country!

I would like to thank also to Saskia, my "Dutch-Mexican friend", for the being one of my best friends.

I cannot be more thankful to Jorge, my husband, and the love of my life and father of my child. Jorge, no tengo palabras para darte las gracias por todo el apoyo que recibi de ti, eres un ganador que no se da por vencido y me haces creer en mi siempre, sin ti, no seria lo que soy ahora. Te amo mi amor. I also thank Ian Marcelo, my child, just for being here. Since you were born I am happier than ever. When I look at you, I know I can do anything; you get the best of me.

I want to thank also in particular to Assiyeh Alizadeh Tabatabai."Carna" you were incredible to me and I love you for all we shared during these four years. I feel really honored that you are my paranymph. I suppose that you already know that you are unofficially Ian's

189

godmother.

To think about all the good moments that I have shared with the amazing people who I was lucky to meet during these four years is just great. To remember dinners, parties, trips, discussions, picnics at the lake, queen's day and many more great and unique experiences.

Maria Pascual, Mulo, Maria Rusca, Elena, Ali, Saul, Ina Krieger, Andreas, Juan Pablo, Paola Ramallo, Julian, Laksmi YOU WERE FAMILY TO ME! I love you guys and I thank you for teaching me and loving me. I am sure that you enjoyed this experience as much as I did, I miss you a LOT!

Finally, I want to thank my father, mother and sister. It does not matter where I will be living in the future; I know that I can always count with you guys. Mama, you are the most positive person that I ever meet, I love you for always encourage me to continue growing. Papa, I always know that I can count on you for anything. Pam, thanks for being here, you changed our lives.

Curriculum vitae

Denys Kristalia Viilla Gómez was born on the 24th of June 1980 in Zacatecas, Mexico. She finished her BSc in Chemical engineer in 2003 and her Msc in Applied sciences with the specialization in environmental engineering in 2006. From 2006 to 2008 she taught courses at the University Autonomous of Zacatecas (UAZ) and at the University Anahuac Cancun in Mexico of environmental engineering, linear algebra and chemistry. She started her PhD Project on 2008 at the Pollution Prevention and Resource Recovery Chair Group of the Department of Enivromental Engineering and Water Technology of UNESCO-IHE, Delft, The Netherlands

List of scientific publications

- Celis L., **Villa-Gómez D.**, Alpuche-Solís A., Ortega-Morales B. and Razo-Flores E. (2009). Characterization of sulfate-reducing bacteria dominated surface communities during start-up of a down-flow fluidized bed reactor. *J. Ind. Microbiol. Biotechnol.* 36(1), 111-121.

- **Villa-Gomez D. K.**, Ababneh H., Papirio S., Rousseau D.P.L., Lens, P.N.L., 2011. Effect of sulfide concentration on the location of the metal precipitates in inversed fluidized bed reactors. *Journal of Hazardous Materials* 192, 200-207.

- **Villa-Gomez D. K.**, Papirio S., van Hullebusch E.D., Farges F., Nikitenko S., Kramer H. and Lens P.N.L. (2012). Influence of sulfide concentration and macronutrients on the characteristics of metal precipitates relevant to metal recovery in bioreactors. *Bioresource Technology* 110 (0):26-34.

- Papirio S., **Villa-Gomez D. K.**, Esposito G., Lens P.N.L. and Pirozzi F. (2012). Acid mine drainage treatment in fluidized bed bioreactors by sulfate-reducing bacteria: a critical review. *Critical Reviews in Environmental Science and Technology* (In press).

- **Villa-Gomez D. K.**, Enright A.M., Listya E., Buttice A., Kramer, H. and Lens P.N.L. (2013) Effect of hydraulic retention time on metal precipitation in sulfate reducing inverse fluidized bed reactors. Submitted to *Journal of Chemical Technology and Biotechnology*.

- **Villa-Gomez D. K.**, Cassidy J., Keesman K., Sampaio R. and Lens P.N.L. Strategies for sulfide control using a pS electrode in sulfate reducing inverse fluidized bed reactors. *Water research*. In press.

- **Villa-Gomez D. K.**, Maestro R., van Hullebusch, E.D., Farges, F., Nikitenko, S., Kramer, H. and Lens, P.N.L. Morphology, mineralogy and solid-liquid phase separation characteristics of Cu and Zn precipitates produced with biogenic sulfide. Submitted to *Environmental Science and Technology*.

- **Villa-Gomez D. K.**, Mushi S., Maestro R., Pakshirajan K. and Lens P.N.L. Influence of process variables and factorial design analysis in an inverse fluidized bed anaerobic reactor towards sulfate reduction at low pH. Submitted to *Journal of Environmnetal Technology*.

SENSE

C E R T I F I C A T E

The Netherlands Research School for the
Socio-Economic and Natural Sciences of the Environment
(SENSE), declares that

Denys Kristalia Villa Gomez

born on 24 June 1980 in Zacatecas, Mexico

has successfully fulfilled all requirements of the
Educational Programme of SENSE.

Delft, 18 October 2013

the Chairman of the SENSE board

Prof. dr. Rik Leemans

the SENSE Director of Education

Dr. Ad van Dommelen

The SENSE Research School declares that Ms. Denys Kristalia Villa Gomez has successfully fulfilled all requirements of the Educational PhD Programme of SENSE with a work load of 47 ECTS, including the following activities:

SENSE PhD Courses
- Environmental Research in Context
- Research Context Activity: Co-organization of the 2nd International Conference "Research Frontiers in Chalcogen Science and Technology", 25-26 May 2010, Delft
- Speciation and Bioavailability

Other PhD Courses
- Innovative technologies for urban wastewater treatment plants
- International School of Crystallization 2009: Foods, Drugs and Agrochemicals
- Project and time management

Management and Didactic Skills Training
- President of the UNESCO-IHE PhD association board
- Co-organisation of the the 3rd International Symposium on Biotechniques for Air Pollution Control
- Supervision of four MSc Theses
- Supervision of an exchange fellow of the Leonardo da Vinci programme

External training at a foreign research institute
- X-ray absorption spectroscopy (XAS) measurements, European Radiation Synchrotron Radiation Facility, France

Oral Presentations
- *Effect of Sulfide Concentration on Local Supersaturation for the Potential Recovery of Metals from Wastewaters* . 1st IWA BeNeLux Young Water Professionals Conference, 30 September - 2 October 2009, Eindhoven, The Netherlands
- *Biogenic Sulfide Production and Selective Metal Precipitation in an Inverse Fluidized Bed Reactor.* 2nd International Conference "Research frontiers in chalcogen cycle science and technology", 25-26 May 2010, Delft, The Netherlands
- *A study of the metal removal mechanisms in sulfate reducing bioreactors: kinetics, speciation and solid phase characterization.* 12th World Congress on Anaerobic Digestion, 31 October - 4 November 2010, Guadalajara, Mexico
- *Metal precipitate characteristics Driving SRB Activity and Community in IFB reactors.* IWA Biofilm 2011 Conference, 27-30 October 2011, Shanghai, China

SENSE Coordinator PhD Education

Drs. Serge Stalpers

T - #0405 - 101024 - C16 - 244/170/12 - PB - 9781138001664 - Gloss Lamination